Shock and Impact
on Structures

International Series on Computational Engineering

Aims:

Computational Engineering has grown in power and diversity in recent years, and for the engineering community the advances are matched by their wider accessibility through modern workstations.

The aim of this series is to provide a clear account of computational methods in engineering analysis and design, dealing with both established methods as well as those currently in a state of rapid development.

The series covers books on the state-of-the-art development in computational engineering and as such comprises several volumes every year covering the latest developments in the application of the methods to different engineering topics. Each volume consists of authored work or edited volumes of several chapters written by the leading researchers in the field. The aim is to provide the fundamental concepts of advances in computational methods as well as outlining the algorithms required to implement the techniques in practical engineering analysis.

The scope of the series covers almost the entire spectrum of engineering analysis. As such, it will cover Stress Analysis, Inelastic Problems, Contact Problems, Fracture Mechanics, Optimization and Design Sensitivity Analysis, Plate and Shell Analysis, Composite Materials, Probabilistic Mechanics, Fluid Mechanics, Groundwater Flow, Hydraulics, Heat Transfer, Geomechanics, Soil Mechanics, Wave Propagation, Acoustics, Electromagnetics, Electrical Problems, Bio-engineering, Knowledge Based Systems and Environmental Modelling.

Professor G.S. Gipson
School of Civil Engineering
Engineering South 207
Oklahoma State University
Stillwater, OK 74078-0327
USA

Professor S. Grilli
The University of Rhode Island
Department of Ocean Engineering
Kingston, RI 02881-0814
USA

Professor M.S. Ingber
University of New Mexico
Department of Mechanical Engineering
Albuquerque
New Mexico 87131
USA

Professor D.B. Ingham
Department of Applied Mathematical Studies
School of Mathematics
The University of Leeds
Leeds LS2 9JT
UK

Professor P. Molinaro
Ente Nazionale per l'Energia Elettrica
Direzione Degli Studi e Ricerche
Centro di Ricerca Idraulica e Strutturale
Via Ornato 90/14
20162 Milano
Italy

Professor Dr. K. Onishi
Department of Mathematics II
Science University of Tokyo
Wakamiya-cho 26
Shinjuku-ku
Tokyo 162
Japan

Professor H. Pina
Instituto Superior Tecnico
Av. Rovisco Pais
1096 Lisboa Codex
Portugal

Professor A. Giorgini
Purdue University
School of Civil Engineering
West Lafayette, IN 47907
USA

Professor W.G. Gray
Department of Civil Engineering and
Geological Sciences
University of Notre Dame
Notre Dame, IN 46556
USA

Dr. S. Hernandez
Department of Mechanical Engineering
University of Zaragoza
Maria de Luna
50015 Zaragoza
Spain

Professor G.D. Manolis
Aristotle University of Thessaloniki
School of Engineering
Department of Civil Engineering
GR-54006, Thessaloniki
Greece

Dr. A.J. Nowak
Silesian Technical University
Institute of Thermal Technology
44-101 Gliwice
Konarskiego 22
Poland

Professor P. Parreira
Departamento de Engenharia Civil
Avenida Rovisco Pais
1096 Lisboa Codex
Portugal

Professor D.P. Rooke
DRA (Aerospace Division)
Materials and Structures Department
R50 Building
RAE Farnborough
Hampshire GU14 GTD
UK

Dr. A.P.S. Selvadurai
Department of Civil Engineering
and Applied Mechanics
McGill University
817 Sherbrooke Street West
Montreal QC
Canada H3A 2K6

Professor P. Skerget
University of Maribor
Faculty of Technical Sciences
YU-62000 Maribor
Smetanova 17
P.O. Box 224
Slovenia

Professor M.D. Trifunac
Department of Civil Engineering, KAP 216D
University of Southern California
Los Angeles, CA 90089-2531
USA

Professor R.P. Shaw
S.U.N.Y. at Buffalo
Department of Civil Engineering
School of Engineering and Applied Sciences
212 Ketter Hall
Buffalo, New York 14260
USA

Dr P.P. Strona
Centro Ricerche Fiat S.C.p.A.
Strada Torino, 50
10043 Orbassano (TO)
Italy

Professor N.G. Zamani
University of Windsor
Department of Mathematics and Statistics
401 Sunset
Windsor
Ontario
Canada N9B 3P4

Thanks are due to J. Martí for the use of Figure 1c, page 267, which appears on the front cover of this book.

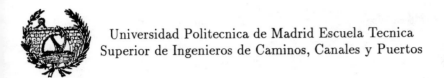

Universidad Politecnica de Madrid Escuela Tecnica
Superior de Ingenieros de Caminos, Canales y Puertos

Shock and Impact
on Structures

Editors:
C.A. Brebbia
V. Sánchez-Gálvez

Computational Mechanics Publications
Southampton, UK and Boston, USA

CMP

C.A. Brebbia
Wessex Institute of Technology
Ashurst Lodge
Ashurst
Southampton SO40 7AA
UK

V. Sanchez-Galvez
ETSI de Caminos, C y P
Universidad Politecnica de Madrid
Ciudad Universitaria
28040 Madrid
Spain

Published by

Computational Mechanics Publications
Ashurst Lodge, Ashurst, Southampton, SO40 7AA, UK
Tel: 44 (0)1703 293223 Fax: 44 (0)1703 292853
Email: CMI@uk.ac.rl.ib.
Intl Email: CMI@ib.rl.ac.uk.

For USA, Canada, and Mexico:

Computational Mechanics Inc
25 Bridge Street, Billerica, MA 01821, USA
Tel: 508 667 5841 Fax: 508 667 7582
Email: CMINA@netcom.com.

British Library Cataloguing-in-Publication Data

A Catalogue record for this book is available
from the British Library

ISBN 1-85312-297-1 Computational Mechanics Publications, Southampton
ISBN 1-56252-221-3 Computational Mechanics Publications, Boston

Library of Congress Catalog Card Number 94-68173

Printed and bound in Great Britain by Bookcraft Ltd, Bath

CONTENTS

PREFACE

This book presents a state of the art of recent developments in shock and impact on structures. It consists of a series of chapters written by renowned scientists at the invitation of the editors.

Chapter 1 deals with the modelling of fast transient loads on structures. It focuses on the constitutive models for material deformation and failure under high rate loading and also in the sources of high rate data. The author, J.A. Zukas, discusses the models available for non-metallic as well as metallic materials.

In Chapter 2 the state of the art for the analysis of blast waves impinging on surface structures is presented. The authors, A. Barbagelata and M. Primavori, describe the way in which blast waves are generated and their interaction with the surfaces. A review is given of the available computer codes for the dynamic analysis of structures subject to blast. In the final section approximate analytical solutions are presented for problems frequently encountered in engineering practice.

In the next chapter N. Jones examines the perforation of metal plates under projectiles with low impact velocity. He presents empirical equations which are suitable for design purposes and investigates the influence of different support conditions.

Chapter 4 by Z. Rosenberg deals with the dynamic response of ceramics to shock wave loading. The new techniques developed to study their behaviour are reviewed, including the modelling of failure criteria as implemented in numerical simulations.

The chapter by A. Miyamoto and M.W. King studies the behaviour of reinforced concrete slabs under soft impact loading. The authors compare layered finite element method results with those obtained in tests on full scale specimens. Their conclusion is that it is possible to predict the ultimate behaviour of these slabs using computer models. Another important conclusion found by these authors is that the impact failure modes for these reinforced concrete structures are related to the loading rate of the impact force-time function. They identify three distinct types of failure modes, i.e. bending failure, intermediate failure (bending to punching shear) and punching shear failure.

C. Ruiz and D. Hughes in Chapter 6 describe impact problems in aeroengines, including the cases of hard and soft impact. Their chapter contributes to the better understanding of how modern aeroengines are able to resist this type of dynamic loading.

Chapter 7 by S. Cescotto and Y.Y. Zhu discusses the use of a type of finite elements to reproduce contact problems, including Coulomb dry friction law and numerical results are presented for dynamic analysis.

The last chapter by J. Martí studies the problem of impact on transport flasks. This is a problem of great importance in highly radioactive materials and the various engineering and licencing requirements on transport and storage of flasks are reviewed.

The book fulfils an important function in providing the most up to date information on structures under shock and impact to the engineering and scientific community.

The Editors

Chapter 1

Numerical simulation of high rate behavior

J.A. Zukas

Computational Mechanics Consultants, Inc., P.O. Box 11314, Baltimore, MD 21239, USA

ABSTRACT

Materials and structures subjected to fast, transient loading (e.g., vehicular impacts, blast and impact loading of structures, the response of containment structures to internal and external stimuli, among others) share several features that make them tractable by wave propagation codes or hydrocodes. The successful simulation of such phenomena requires adequate spatial resolution, a constitutive model that incorporates the features observed experimentally and data for the constitutive model appropriate to the range of strain rates encountered in reality. This chapter focuses on the available constitutive models for material deformation and failure under high rate loading and also sources of high rate data. It is found that for metallic materials, adequate models and data exist to model large deformation effects quite accurately. However, failure models range from inadequate to impractical and the available data for existing models is very limited. For non-metallic materials, both definitive computational models and an adequate data base for high rate loading simulations are lacking. Much of the current work in this area relies on ad hoc models, some of which lack a firm physical foundation.

BLAST vs. IMPACT LOADING

Explosions result from the sudden release of energy. The energy may come from an explosive, wheat flour dust in a grain elevator, pressurized steam in a boiler or an uncontrolled nuclear transformation. The accumulated energy is dissipated in various ways such as in blast waves, the propulsion of missiles or by thermal or ionizing radiation (Kinney and Graham [1]).

As the blast wave travels away from its source, the overpressure (pressure in excess of atmospheric pressure) at the front steadily decreases and the pressure behind the front falls off in a regular manner. After a short time, the pressure behind the front drops below atmospheric so that an underpressure rather than overpressure exists. During this negative (rarefaction or suction) phase, a partial vacuum is produced. Air is sucked in instead of being pushed away. At the end of the negative phase, which is slightly longer than the positive phase, the pressure returns to ambient. Underpressures usually have a magnitude of less than 28 kPa whereas overpressure can exceed 2 MPa.

Structures suffer damage from air blast when overpressures exceed 3.5 kPa. The distance to which the overpressure level extends depends on the energy yield of the explosion. The blast loading is seen as a lateral dynamic pressure, applied rapidly, lasting for a second or more and continuously decreasing in strength. Factors that most strongly affect the structural response are the inertia of the structure, as measured by the mass and strength of the structure, the overall structural design and the ductility of the materials and members comprising the structure. The strength of the structure is not only a material characteristic but includes the massiveness of the construction and redundancy of supports. If the strength is not isotropic, then the orientation of the structure with respect to the burst becomes important (Kinney and Graham [1]; Glasstone and Dolan [2]; Bailey and Murray [3]).

Impact, by contrast, is a localized phenomenon. Upon collision with a target, the process of transforming some or all of the kinetic energy of the striker into deformation and failure of the colliding materials is governed by wave propagation. On contact, a compression wave propagates into both striker and target with an initial intensity of $\rho c v$, where ρ represents the density, c the sound speed, a material characteristic, and v the particle velocity. Because of geometric effects, this initial intensity rapidly decays to a value of $v^2/2$. The presence of free surfaces and material interfaces causes multiple reflections of the propagating waves causing the occurrence of sharp compressive and tensile stress gradients. Depending on the duration of the tensile pulses and their intensity, material failure in the form of physical separation of material may occur by a number of mechanism (see Curran, Seaman and Shockey [4]; Backman and Goldsmith [5]; Johnson [6]; Zukas et al. [7]; Zukas [8,9]; Bushman et al. [10]; Blazynski [11] for comprehensive discussions on impact phenomena).

The stress waves in the colliding solid propagate at a characteristic velocity that is a property of the material. The particle velocity (the speed at which material particles are actually displaced in the solid), however, governs the ensuing deformation. Because of this, coupled with the presence of multiple wave reverberations due to geometric effects, the most intense deformations occur within

3-6 characteristic striker dimensions (3-6 diameters if the striker has regular geometry such as a cylinder, sphere or cone). Within this region, strain rates in excess of 100000/s are not uncommon, as are plastic strains exceeding 60% and pressures well in excess of the material strength. Loading and response times are in the submillisecond regime. Unlike structural response problems, the high frequency components of the structure are excited and lead to material failure by a wide variety of mechanisms.

Because of the localized nature of the impact processes (Wright and Frank [12]), the geometries that need to be considered can be somewhat simplified and are shown in Figure 1. Unlike the structural dynamics problem, catastrophic failure by penetration or perforation can occur long before the distant boundaries have any effect. Thus, boundary conditions tend to be very simple, allowing either for perfect transmission or reflection of the incoming signal. Initial conditions most often specify the velocity of the striker, although these can be augmented by surface pressures or shears depending on the problem.

When it becomes necessary to consider the lethal effects of debris caused by blast loading, the blast and impact problems share several common features. Wave propagation effects must be accounted for explicitly since the interaction of stress waves with free surfaces, material interfaces and geometric discontinuities lead to material failure such as spallation or fragmentation. Inertia effects must be accounted for in considering both types of problems. This implies the use of explicit time integration techniques numerical analyses as well as contact-impact (sliding interface) logic to properly account for momentum distribution between flying debris and surrounding structures. Both types of problems need constitutive models for deformation and failure at high strain rates. The "material data" or constants that are embedded in these models must also be determined from wave propagation experiments, i.e., at strain rates appropriate to the problems being analyzed. Hence, the early stages of a problem in blast loading and most high velocity impact problems can be successfully analyzed by wave propagation codes, often referred to as hydrocodes.

NUMERICAL METHODS

Numerical techniques for fast, transient loading and associated computer codes have received considerable attention in the literature lately and need not be considered in detail here. See the reviews by Zukas [7,9], Anderson and Bodner [13] and Belytschko and Hughes [14] for details. The key point is that the accuracy of any calculation involving fast, transient loading depends on three major factors: the computational grid, the constitutive model and the data used in the constitutive model. Gridding considerations have been discussed in some detail by Creighton [15], Zukas [16] and Johnson and Schonhardt [17], among others. Here we shall concentrate on current developments in constitutive modeling and sources of high

TYPES OF IMPACT PROBLEMS

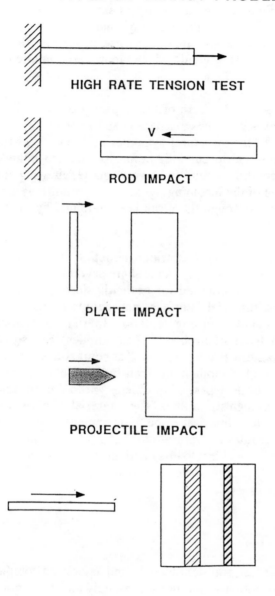

Figure 1: Typical Geometries for Numerical Simulation of Impact

strain rate data for wave propagation calculations.

Strength Models

Modern hydrocodes decouple the material response for metallic materials into a volumetric and a deviatoric part. The volumetric or hydrodynamic response is obtained through an equation of state. By far the most popular equation of state (EOS) in existing codes is the Mie-Gruneisen equation of state, which makes use of Hugoniot data obtained from plate impact experiments and includes an energy term to reach off-Hugoniot states. The Mie-Gruneisen EOS assumes that the material will respond as a solid throughout its loading regime. For transitions from solid to vapor or liquid states, another equation must be used. The Tillotson EOS has been very popular for many years for problems in hypervelocity impact where such transitions occur with high frequency. Examples of equations of state popular in modern hydrocodes are shown in Zukas [16]. Extensive compilations of equation of state data exist, due to Van Thiel [18], Marsh [19], Dobratz [20], Kohn [21] and others. For situations where EOS data does not exist, there are at least a dozen methods by which it can be obtained. Thus, years of experimentation and systematic collection of shock wave data have made computation of the hydrodynamic response of materials almost a routine matter. Such is not the case when considering material strength, or deviatoric, effects.

An incremental elastic-plastic formulation is used to describe the shear response of metals in present finite-difference and finite-element codes. The plasticity descriptions are based on the assumed decomposition of the velocity strain tensor into elastic and plastic parts together with incompressibility of the plastic part. The von Mises yield criterion is typically used to describe the onset of yielding. Provision is made to account for strain hardening, compressibility and thermal effects but more often than not such data is not available for practical calculations. Earlier hydrocodes incorporated the Jaumann stress rate. Because of difficulties with these formulations at strains exceeding 40% (see Walters and Zukas [22]), the Jaumann rate is gradually being replaced with alternatives (Green and Naghdi [23], Dienes [24]).

The above formulations are virtually common to all wave propagation codes. In addition, some codes also allow for a number of alternative material descriptions: elastic (isotropic or orthotropic); piecewise linear elastic-plastic models with strain hardening effects; linear viscoelastic; thermo-elastic; thermo-plastic with material temperature-dependent properties. In addition to the von Mises flow rule, general isotropic, kinematic and combined strain hardening rules allow yield surface translation as well as expansion or contraction.

For non-metallic materials a variety of models, some ad hoc, some with theoretical basis are available. NIKE, for example, uses a p-alpha model (described below) to model soils and crushable materials. ADINA includes a concrete model with multi-axial material failure envelopes on tension cracking and compression crushing and strain softening; a Drucker-Prager type model for soils and rocks with tension cutoff and compression cap; and a curve (empirical) description soil and rock model with tension cutoff or cracking.

Porous materials are typically handled through a p-alpha type model first developed by Herrmann [25]. Data for this model must be determined from individual materials from a number of independent experiments. The generic form of such a model is depicted in Figure 2.

Geological materials are commonly handled through a CAP model (Gupta and Seaman [26]; Wright and Baron [27]), shown in Figure 3. This has also been adapted for porous materials, concrete and ceramics by judicious adjustment of the various parameters. A non-associative flow rule is commonly used to describe the shear envelope. The cap moves depending on the degree of volumetric response.

Dynamic events often involve increases in temperature due to adiabatic heating. To accurately predict the response of a material, the effect of temperature on the flow stress must be included in a constitutive model. An attempt to account for the effects of strain, strain rate and temperature is made in the model proposed by Johnson and Cook [28-29]. This is an empirical model involving five constants A,B,C,n and m. The functional form of the model constitutes the authors' best guess as to how material behaves under high rate loading. It is not based on any theory. The constants must be determined from experiments (Lips et al. [30]; Johnson and Holmquist [31]; Rajendran [32]). Both the experiments and the model decouple the effects of the parameters which enter it. One result is the implication of the model that strain rate sensitivity is independent of temperature, a feature that is not generally observed for most metals (Nicholas and Rajendran [33]). In fact, rate sensitivity is found to increase with increasing temperature while flow stress decreases. The Johnson-Cook strength and failure models are shown in Table 1. Parameters for the strength model are in Table 2 (Nicholas and Rajendran [68]).

Zerilli and Armstrong [34] used dislocation dynamics concepts which also accounts for strain, strain rate and temperature effects in a coupled manner. For f.c.c. metals their model takes the form

$$\sigma = C_0 + C_2 \epsilon^n [e^{(-C_3 T + C_4 T \ln \dot{\epsilon})}]$$

while for b.c.c. metals it is given as

$$\sigma = C_0 + C_1 [e^{-C_3 T + C_4 T \ln \dot{\epsilon}}] + C_5 \epsilon^n$$

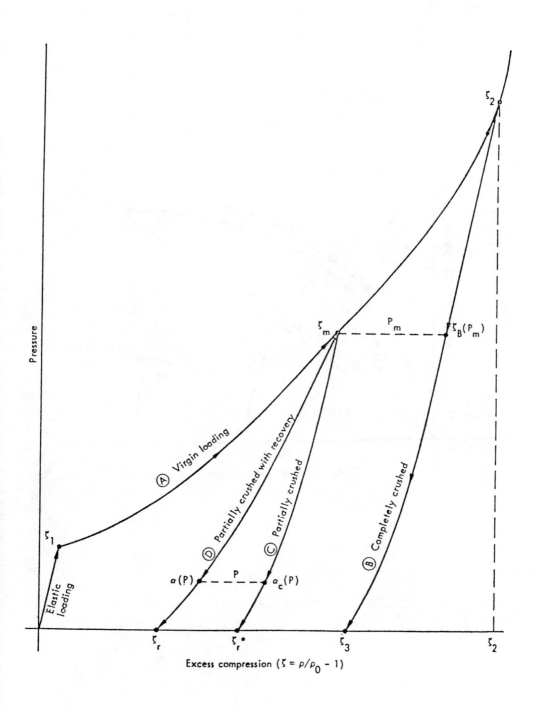

Figure 2: Constitutive Model for Porous Media

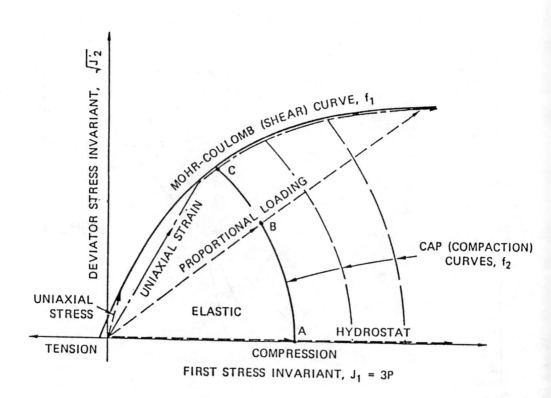

Figure 3: CAP Model

Table 1. Johnson-Cook Strength and Fracture Models

STRENGTH MODEL FOR VON MISES TENSILE FLOW STRESS

$$\sigma = [A + B\epsilon^n][1 + C\ln\dot{\epsilon}^*][1 - T^{*m}]$$

STRAIN — STRAIN RATE — TEMPERATURE

WHERE A, B, n, C, m ARE MATERIAL CONSTANTS

FRACTURE MODEL

$$D = \sum \frac{\Delta\epsilon}{\epsilon^f} \quad \text{(DAMAGE)}$$

$$\epsilon^f = [D_1 + D_2 \exp D_3\sigma^*][1 + D_4\ln\dot{\epsilon}^*][1 + D_5 T^*] \quad \text{(FRACTURE STRAIN)}$$

PRESSURE — STRAIN RATE — TEMPERATURE

$\sigma^* = \sigma_m/\bar{\sigma}$

WHERE $D_1.....D_5$ ARE MATERIAL CONSTANTS

$D \geq 1.0$ GIVES FRACTURE

Table 2. Parameters for Johnson-Cook Strength Models

STRENGTH MODEL CONSTANTS FOR JOHNSON-COOK MODEL

MATERIAL	C_1 MPa	C_2 MPa	C_3	n	m	REMARKS
OFHC Copper	89.63	291.64	0.025	0.31	1.09	800°F; Anneal/60 min.
Cartridge Brass	111.69	504.69	0.009	0.42	1.68	1000°F, Anneal/60 min.
Nickel 200	163.40	648.10	0.006	0.33	1.44	1300°F, Anneal/30 min.
Armco Iron	175.12	3799.90	0.06	0.32	0.55	1700°F, Anneal/60 min.
Carpenter Electric Iron	289.58	338.53	0.055	0.40	0.55	1000°F, Anneal/60 min.
1006 Steel	350.25	275.00	0.022	0.36	1.00	
2024-T351 Aluminum	264.75	426.09	0.015	0.34	1.00	
7039 Aluminum	336.46	342.66	0.01	0.41	1.00	
4340 Steel	792.19	509.51	0.014	0.26	1.03	
S-7 Tool Steel	1538.89	476.42	0.012	0.18	1.00	
Tungsten	1505.79	176.50	0.016	0.12	1.00	7% Ni, 3% Fe
Depleted Uranium	1079.01	1119.69	0.007	0.25	1.00	0.75 Ti
Tantalum	140.00	300.00		0.30	0.70	[ref. Lips et al. (1987)]

*not provided

Table 3. Bodner-Partom Model

$$\dot{\epsilon}_{ij}^{p} = D_0 \, e^{\left(-\frac{(n+1)}{n} \left(\frac{Z^2}{3 J_2}\right)\right)^n} \frac{S_{ij}}{\sqrt{J_2}}$$

$$\dot{Z} = m \, (Z_1 - Z) \, \dot{W}_p$$

D_0 is the limiting strain rate, n is strain rate sensitivity parameter
m is strain hardening parameter
Z_0 and Z_1 are initial and saturated values of the internal state variables Z
Z describes resistance to plastic flow, and loading history dependency.
n is a temperature dependent constant. W_p is plastic work.

Table 4. Bodner-Partom Model Constants

Material	Z_0 (GPa)	Z_1 (GPa)	n	m_0 GPa^{-1}	m_1 GPa^{-1}	α GPa^{-1}	A	B
C1008 Steel	5.5	7.0	0.4	15	0	0	0.245	46
HY100 Steel	2.4	3.6	1.2	10	0	0	NA	NA
1020 Steel	0.64	0.93	4.0	30	0	0	NA	NA
MAR-200 Steel	2.2	2.4	4.0	5	0	0	NA	NA
Armco Iron	2.65	4.2	0.58	56	0	0	NA	NA
OFHC Copper	0.8	6.6	0.4	11	150	1500	NA	NA
6061-T6 Aluminum	0.45	0.55	4.0	120	0	0	-2.86	2343
7039-T64 Aluminum	0.56	0.76	4.0	28	0	0	NA	NA
Pure Tantalum	1.3	3.1	0.74	20	0	0	NA	NA
W-2 Tungsten	8.75	10.0	0.58	150	0	0	0.166	134
Nickel 200	0.32	0.82	4.0	40	0	0	NA	NA
MAR-250 Steel	2.5	2.7	5.0	20	0	0	NA	NA
AF1410 Steel	2.4	2.75	5.0	15	0	0	NA	NA

NA -- The high temperature constants are "Not Available"

Here, Y is the strength, T the temperature, the C's are material constants,   is the plastic strain and the plastic strain rate.   The rationale for the difference in the two forms is in the stronger dependence of yield stress on temperature and strain rate is known to result for b.c.c. metals than f.c.c. metals. The constants for the model have been determined for only two materials. A comparison was made between the Johnson-Cook and Zerilli-Armstrong models for Taylor cylinder impact experiments (Johnson and Holmquist [31]). The Zerilli-Armstrong model gave slightly better correlation with experimental results although, being small strain models, neither was very accurate in describing large strain behavior (Nicholas and Rajendran [33]).

The Bodner-Partom model [35] is based on dislocation dynamics concepts and treats strain rate and temperature effects in a coupled manner to more realistically account for observed behavior. It has been used in analysis of rate history effects (Bodner and Merzer [36]), propagation of plastic waves in long rods (Bodner and Aboudi [37]) and precursor decay analysis (Nicholas, Rajendran and Grove [38-39]). The Bodner-Partom Model is shown in Table 3 while Table 4 lists available constants. Procedures for determining the constants are given by Rajendran [32,40].

Models developed by Steinberg, Cochran and Guinan [41] and Steinberg [42-43], assume strain rate saturation and allow variation of shear modulus and yield strength with pressure and temperature:

$$G = G_0 \left[1 + \left[\frac{G'_p}{G_0}\right]\frac{p}{\eta^{1/3}} + \left[\frac{G'_T}{G_0}\right](T - 300)\right]$$

$$Y = Y_0(1 + \beta\epsilon)^n \left[1 + \left[\frac{Y'_p}{Y_0}\right]\frac{p}{\eta^{1/3}} + \left[\frac{G'_T}{G_0}\right](T - 300)\right]$$

G and Y are the degraded shear modulus and strength, respectively, G_0 and Y_0 are the same quantities at the reference state. The primed quantities represent partial derivatives with respect to pressure and temperature, η is the ratio of current to initial specific volume and β and n are model parameters. Steinberg [43] catalogs the available parameters for the model.

Other models which account for the variation of strength with strain, strain rate, temperature, pressure or other parameters are discussed in the literature (Nicholas and Rajendran [33]; Zerilli and Armstrong [34]; Rajendran [32]; Follansbee and Kocks [44]; Meyer [45]). These are generally one-dimensional models and have been used to interpret various experiments involving high rate behavior. In principle, one-dimensional models can be generalized to three dimensions through the use of effective stress and strain formulations. Except for isolated incidences to support specific laboratory experiments, these have not found their way into production computer codes and will not be mentioned further.

In addition, a large number of ad hoc models exist for solids, liquids and explosives. These usually lack theoretical foundation and are little more than attempts to fit existing and incomplete data bases. There have been as many successful calculations performed with such models as unsuccessful ones. The current interest is in computational models for ceramics and for reinforced concrete. At present, no definitive computational models have emerged. However, this is an area of active research.

Failure Models

Failure models for structural dynamics and wave propagation codes have been discussed extensively by Woodward [46], Nicholas and Rajendran [33], Rajendran [32], and Zukas [8], among others. In many structural impact calculations explicit failure models are not needed. The criterion for acceptable performance is often based on the degree of plastic deformation of a critical member or members. It can be the absence of buckling, the load seen at a particular location, the effectiveness of energy-absorbing construction and a variety of other criteria as required for safety or specified by regulations for specific applications. Wave propagation can result in failure which amounts to actual separation of material by a variety of mechanisms (spallation, tearing, ductile or brittle fracture, adiabatic shear, etc.). These criteria fall into three categories:

(a) Instantaneous - these are based on maxima of field variables such as effective plastic stress, effective plastic strain, principal stress or strain, plastic work, internal energy, maximum tensile pressure or some combination of these. Once the criterion is met in the calculation, the stresses associated with a given cell or element are immediately zeroed out and the cell or element can no longer carry any type of load. A modification of this is to limit the load-carrying capability of an element to compressive loads only until another level of the failure criterion is reached whereupon the cell or element is said to have failed totally and is ignored in further computations.

The simplest, and one of the earliest, instantaneous failure models was the pressure cutoff. When the hydrodynamic pressure reaches a critical value in tension, failure is assumed to occur instantaneously. The pressure is not allowed to grow beyond this user-specified critical value and further expansion occurs at this value of pressure. Any shear stresses in a computational cell or element are set to zero.

Recompression is permitted and if this occurs the computational cell can again carry compressive pressures. Related to this are the volumetric strain and maximum distension criteria. These were used in the early finite-difference codes to study hypervelocity impact where hydrodynamic pressures dominate. They were useful since for the short loading and response times governing hypervelocity impact or shock loading situations, micromechanical considerations can be ignored as a first order approximation.

For situations where pressures are on the order of material strength for the bulk of the response time, initiation criteria based on some measure of stress or strain have been more effective. The post-failure response has been modeled in a number of different ways. In the simplest approach, the cell or element loses its load-carrying capability once the failure criterion is satisfied. In a variant of this, stresses may be relaxed to zero over several computational cycles. An approach now incorporated in a number of Lagrangian wave propagation codes is the introduction of discrete free surfaces along failure planes to model events such as spall, fragmentation or plugging. With this approach, void opening and closing is automatically taken into account (Poth et al. [47-48]; Ringers [49]). Another approach assumes that the failure plane is normal to the maximum principal direction. The stress tensor calculated for undamaged material is then modified to account for the fact that the failure plane has been weakened (Bertholf et al. [50]). Compressive stresses may be transmitted if the opening subsequently closes.

Some situations cannot be effectively addressed by models based on a single variable. Successful use has been made of the so-called P/Y model where P represents hydrodynamic pressure, determined from an equation of state, and Y the flow stress. This is based on the work of Hancock and MacKenzie [51] who showed that the ductility of steel depends markedly on the triaxiality of the state of stress existing within the steel. The high hydrodynamic pressure present in many fast, transient loading situations can have a strong effect on the strain to fracture. Combining the effective stress with the hydrodynamic pressure provides a single non-dimensional parameter which gives a measure of the stress state triaxiality. In this model, the plastic strain and state of stress, as characterized by P/Y, become the field variables characterizing fracture (Figure 4).

The determination of P/Y versus effective plastic   strain curves is discussed by Nash and Cullis [52] and Matuska and Osborn [53]. A typical example for rolled homogeneous armor is shown in Figure 5. The criterion has been incorporated in the HULL code. Excellent results have been obtained by Nash and Cullis [52] and Matuska and Osborn [53] in the few cases where it has been applied to high velocity impact problems.

Since material failure has long been known to be a time-dependent process (e.g., Oscarson and Graff [54]), why rely on such simple criteria? There are a number of reasons. Often, accounting for material failure in even the simplest ways produces

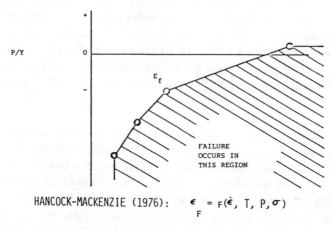

HANCOCK-MACKENZIE (1976): $\epsilon_F = F(\dot{\epsilon}, T, P, \sigma)$

HIGH STRENGTH MATERIAL:

$$\epsilon_F = \alpha \text{ EXP} \left(\frac{-3P}{2\bar{\sigma}}\right)$$

SUBSTANTIAL PLASTIC FLOW BEFORE VOID NUCLEATION:

$$\epsilon_F = e_N + \alpha \text{ EXP} \left(\frac{-3P}{2\bar{\sigma}}\right)$$

e_N = VOID NUCLEATION STRAIN

ϵ_F = MAXIMUM PRINCIPAL TENSILE STRAIN

Figure 4: Hancock-MacKenzie (P/Y) Failure Model

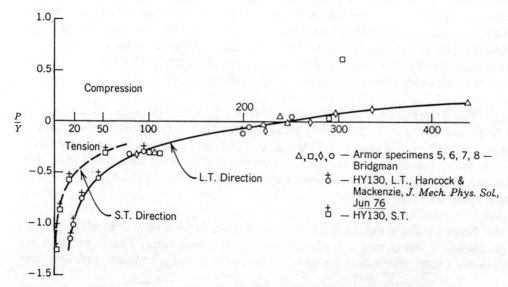

Figure 5: Hancock-MacKenzie Model Data for Rolled Homogeneous Armor

results that show good correlation with experiments. As loading rate increases, there is progressively less and less time for micromechanical mechanisms to be activated so that for situations such as hypervelocity impact a simple maximum tensile stress or stress gradient criterion is sufficient to correctly reproduce the location of the spall plane and the thickness and momentum of spall fragments. The simple criteria have the further advantage that they are easy to include in code calculations and the one or two parameters needed for a failure criterion can be readily estimated from existing data or obtained from a simple laboratory experiment. The principal disadvantage of the instantaneous criteria is that they are not physically correct. Other drawbacks are that they are independent of stress state or loading history, do not allow the possibility of partial failure and cannot be used to model degradation of the material as damage progresses.

(b) Time-dependent - these include a time history effect and consider not only the amplitude of the stress but the duration of the stress pulse as well. Damage is allowed to accumulate until a critical value of the damage parameter is reached, whereupon abrupt failure occurs.

One of the first to be used to study spallation in plate impact experiments was the Tuler-Butcher [55] criterion:

$$K = \int_0^t (\sigma - \sigma_0)^{\lambda} dt$$

The criterion assumes the existence of a stress level below which no failure occurs and allows for accumulation of damage until a critical value K is reached. It has three adjustable parameters and represents a tractable problem in material characterization, though more complex and expensive than determining a single parameter. Mescall and Papirno [56] found better correlation with HEMP code analyses and spall experiments using the Tuler-Butcher criterion than with instantaneous criteria.

A cumulative damage model due to Johnson [57] has been incorporated in a number of codes including EPIC and CTH (Table 1). The parameters D_1 - D_5 must be determined experimentally. Note that the first term in the brackets is equivalent to the Hancock-MacKenzie criterion. The model has been tested for a limited number of cases and found to provide qualitative agreement with experiments. Reasons stated for the discrepancy between computations and experiments include uncertainties in the fracture characteristics of the materials studied (OFHC Copper, ARMCO Iron); failure of extrapolation of model predictions to regions of strain rate, temperature and pressure encountered in high velocity impact situations, and; the possibility that damage does not accumulate as stated in the model.

Cumulative damage models have the advantage of being more realistic since, in contrast to instantaneous models, failure now depends on a critical level of stress acting over a finite time. They are advantageous from a computational standpoint since, lacking the very strong dependence on peak stress of the instantaneous failure models, good results are possible with relatively large computational cells. Moreover, data from experiments with simple pulse shapes can readily be extended to treat more general shapes (Oscarson and Graff [54]). However, the degree of material characterization is increased. Several well-designed experiments must be performed to determine the three or more parameters of a time-dependent damage model. In the process a certain amount of uncoupling of relevant phenomena occurs. The calibrating experiments are of necessity one-dimensional and focus on only one aspect (large strain, strain rate, temperature or pressure) of the problem whereas the models are used to simulate three-dimensional impact effects where strain rate, pressure and temperature effects occur simultaneously, usually at loading rates considerably higher than those of the calibrating experiments for such models. Furthermore, some of the quantities required by these models cannot be measured directly and must be inferred from a combination of experiments and computer simulations. Thus, there is a significant increase in facilities and labor requirements to develop data for these models, in contrast to the single parameter required for instantaneous models which can often be inferred from existing data.

(c) Micromechanical - these take into account the microstructural evolution of the damage; they are complex but have the advantage of following accurately the physical process that leads to failure of materials.

Little else will be said here about such models since, by and large, they have not made their way into production codes. The main problem is not their complexity but the lack of data for such models for structural materials of general interest. Excellent results have been obtained for one-dimensional simulations of spall and adiabatic shearing failure. However, full characterization of a model accounting for void formation, nucleation, growth and coalescence requires a large number of experiments and investment of several man-years just to characterize one material. The cost is prohibitive for most practical applications. See Curran [58], Nicholas and Rajendran [33] and Zukas [8-9] for details.

A micromechanically-based model for ductile failure that employs but five parameters and has successfully modeled wave propagation and spall in plate impact experiments is due to Rajendran et al. [59-62]. The model accounts for damage nucleation and growth and coalescence. The Bodner-Partom law is used as the constitutive relation for the void-free material and a model is included for the aggregate. Procedures for calibrating the model and model parameters for copper, tantalum and a variety of steels are provided by Rajendran [33].

In addition to the above, models based on mechanical energy and on the tensile stress gradient have also been used for specific applications. Noteworthy is the stress gradient model used by Breed, Mader and Venable [63] to calculate spall failure with the one-dimensional SIN code (Mader [64]). The calculations predicted not only the number of spall fragments but their positions as well, in good agreement with experimental data in Mader, Neal and Dick [65].

In practice, 99% of all calculations are performed with instantaneous failure models, even though it is well known that both cumulative and micromechanical damage models are physically more realistic. There are two major reasons for this. Often, the simple models produce reasonable results, especially if the only data available for comparison with calculations are global (deformations, residual masses, velocities, hole sizes, energies). But principally, the simple criteria are used for lack of data for the more complex models for materials of practical interest.

MATERIAL DATA

The primary source of high strain rate material data for computations comes from split-Hopkinson bar experiments. Complete descriptions of split-Hopkinson bar experiments in compression, tension and torsion are given by Nicholas [66] and Nicholas and Rajendran [33,68]. The apparatus is shown schematically in Figure 6. The variation of strain rate during the test is depicted in Figure 7. By measuring the transmitted and reflected strain pulses in the bar, the stress, strain and strain rate in the specimen can be determined from

$$\epsilon_s = -\frac{2c_0}{L}\int_0^t \epsilon_r \, dt$$

$$\sigma_s = E \left[\frac{A}{A_s}\right]\epsilon_t$$

$$\dot{\epsilon}_s = -\frac{2c_0}{L}\epsilon_r$$

where the subscripts s,r and t refer to the test specimen, the transmitted pulse and

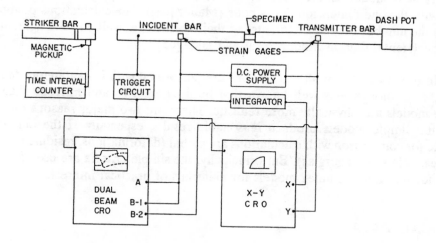

Figure 6: Schematic and Instrumentation for Compression split-Hopkinson Bar

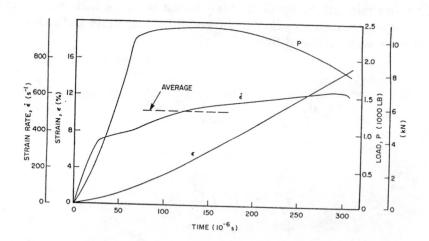

Figure 7: Tensile Hopkinson Bar Data for Ti 6Al-4V Showing Variation in Strain, Strain Rate and Specimen Loading During Test

the reflected pulse, respectively. A represents the cross-sectional area of the bar and specimen, as labeled, L is the specimen length and c the elastic sound speed in the bar. For this analysis to hold, the incident and transmitter bars must remain elastic and a large number of wave reverberations must have occurred in the sample to justify the assumption that the average forces are equal at both ends of the specimen. Extensive data from split-Hopkinson experiments appear in Nicholas [67] and Nicholas and Rajendran [68].

The so-called Taylor Cylinder Test is a technique for determining the dynamic yield stress of a material first developed by Taylor and Wiffin (Nicholas [66]; Johnson [6]). A right-circular cylinder is impacted against a "rigid" target and measurements are made of the deformed shape. These parameters are then used in a simple analysis first formulated by Taylor using conservation of momentum principles to yield and average dynamic yield strength. The analysis for the Taylor Cylinder test was critically reviewed by Hawkyard and an alternate derivation based on energy conservation developed which provided improved correlation at higher striking velocities. Details are given in the book by Johnson [6].

Very elegant experiments involving bar-bar impacts were performed by Bell [69]. The experimental setup is shown in Figure 8. Through measurement of the local surface deformation and assumption of the rate independence of dynamic plasticity, stress-strain curves could be constructed from the experiments at strain rates approaching 10000/s.

The above measure dynamic behavior in primarily uniaxial stress conditions. Plate-plate impact experiments generate state of uniaxial strain and thus can attain very high pressures in comparison to material strength. These are used primarily to study shock wave propagation and failure due to shock loading in materials. Both the amplitude and duration of the stress pulse can be controlled in these experiments by altering the thickness of the impacting plate (duration) and its velocity (stress amplitude). Measurements are made indirectly by recording rear surface velocities and deflections.

A novel approach which combines the best of uniaxial stress and strain experiments has been developed by Clifton [70-72] and his associates at Brown University. A schematic of their pressure-shear experiment is shown in Figure 9. The test generates compression and shear waves resulting in shear strain rates exceeding 10000/s together with superimposed normal pressures (Figure 10). The technique is still new and available data limited. It promises, however, to answer many questions regarding the behavior of materials at strain rates in excess of 10000/s.

Other techniques are discussed in the literature (e.g. Nicholas [66] and Nicholas and Rajendran [33,68]).The above, however, are the primary sources of high rate material behavior information for those engaged in computational efforts. Unfortunately, there are no compilations of such data similar to the cataloging of

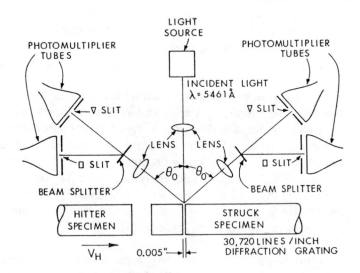

Figure 8: Schematic of Bell Bar Impact Experiment

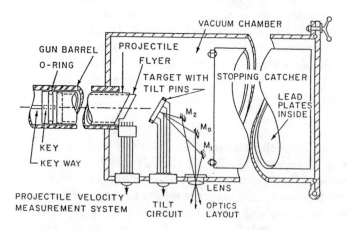

Figure 9: Schematic of Pressure-Shear Experiment

PRESSURE SHEAR EXPERIMENT

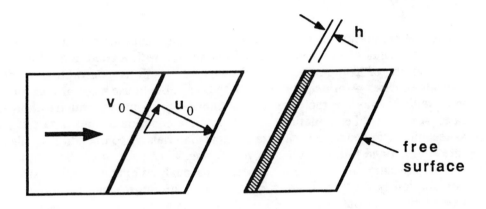

NORMAL AND SHEAR STRAIN RATES IN SPECIMEN:

$$\dot{\varepsilon} = \frac{u_0 - u_{fs}}{h}$$

$$\dot{\gamma} = \frac{v_0 - v_{fs}}{h}$$

NORMAL AND SHEAR TRACTIONS ON SPECIMEN:

$$\sigma = \frac{1}{2}\rho c_1 u_{fs}$$ c_1 = longitudinal wave speed

$$\tau = \frac{1}{2}\rho c_2 u_{fs}$$ c_2 = shear wave speed

Figure 10: Pressure - Shear Experiment

equation of state data that has gone on consistently for the last thirty years. The data is scattered over a wide literature and requires expenditure of many manhours merely to determine its existence for a given material. Worse, much of the data resides in private files and has never been published.

SUMMARY

Numerical simulation of the behavior of metallic materials and structures to fast, transient loading can be accomplished readily and with a good deal of accuracy. A wide variety of computational tools exist, as well as constitutive models for specialized cases. A principal drawback to being able to use wave propagation codes in a predictive mode is the poor state of knowledge of material failure at high strain rates. A number of compilations of high pressure equation of state data exist, as well as techniques to generate such data for new materials. A compendium of high strain rate strength and failure data is as yet lacking. This places great hardship on those who require such information for experiments, analyses or computations since it must be laboriously dug up from a vast, diverse literature and private collections. Considering that only some 20% of the cost of a computational exercise is direct computer costs and the remaining 80% is expended on labor costs, the lack of such a compilation represents a terrible and unnecessary waste.

For nonmetals, the picture is much worse. In many cases, definitive computational models do not exist. Equally lacking is equation of state and high rate strength and failure data, especially for materials of such practical interest as concrete, various composites and ceramics. Many calculations are performed with ad hoc models, some of which lack a firm physical foundation. It is in this area where considerable research is required, both to devise appropriate test techniques to measure high rate response and to develop the appropriate constitutive models.

REFERENCES

1.   Kinney, G.F. and Graham, K.J.,Explosive Shocks in Air, 2nd ed.,Springer-Verlag, New York

2.   Glasstone, S. and Dolan, P.J., The Effects of Nuclear Weapons, 3rd ed.,U.S. Government Printing Office, Washington, D.C., 1977

3.   Bailey, A. and Murray, S.G., Explosives, Propellants and Pyrotechnics, Brassey's (UK), London, 1989

4.   Curran, D.R., Seaman, L. and Shockey, D.A., Dynamic Failure in Solids, Physics Today, Vol. 30, pp. 46-55, 1977

5.   Backman, M.E. and Goldsmith, W., The Mechanics of Penetration of Projectiles Into Targets, Int. J. Engng. Sci., Vol. 16, pp.1-99, 1978

6.   Johnson, W., Impact Strength of Materials, Edward Arnold, London, 1972

7.   Zukas, J.A., Nicholas, T., Swift, H.F., Greszczuk, L.B. and Curran, D.R., Impact Dynamics, Wiley

Interscience, New York, 1982. Republished by Krieger Publishing Co., Malabar, FL, 1992

8. Zukas, J.A., Fracture With Stress Waves, Chapter 6, Materials At High Strain Rates, (Ed. Blazynski, T.), pp.219-242, Elsevier Applied Science, London, 1987

9. Zukas, J.A. (Ed.), High Velocity Impact Dynamics, Wiley Interscience, New York, 1990

10. Blazynski, T. (Ed.), Materials at High Strain Rates, Elsevier Applied Science, London, 1987

11. Bushman, A.V., Kanel', G.I., Ni, A.L. and Fortov, V.E., Intense Dynamic Loading of Condensed Matter, Taylor and Francis, Washington, DC, 1993

12. Wright, T.W. and Frank, K., Approaches to Penetration Problems, in Impact: Effect of Fast, Transient Loading (Eds. Amman, W., Liu, W.K., Studer, J.A. and Zimmerman, T.), A.A. Balkema, Rotterdam, 1988

13. Anderson, C.E., Jr. and Bodner, S.R., Ballistic Impact: The Status of Analytical and Numerical Modeling, Int. J. Impact Engng., Vol. 7, #1, pp.9-35, 1988

14. Belytschko, T. and Hughes, T.J.R., Computational Methods for Transient Analysis, North-Holland, Amsterdam, 1983

15. Creighton, B.M., Numerical Resolution Calculation for Elastic-Plastic Impact Problems, Memorandum Report BRL-MR-3418, U.S. Army Ballistic Research Laboratory, December 1984

16. Zukas, J.A., Some Common Problems in the Numerical Modeling of Impact Phenomena, Comput. Sys. Engng., Vol.4, #1, pp. 43-58, 1993

17. Johnson, G.R. and Schonhardt, J.A., Some Parametric Sensitivity Analyses for High Velocity Impact Computations, Nuc. Eng. Des., Vol. 138, pp. 75-91, 1992

18. van Thiel, M., Equation of State Data, Report UCRL-50108, Vols. 1-3, Lawrence Livermore National Laboratory, 1977

19. Marsh, S.P. (Ed.), LASL Shock Hugoniot Data, U. of California Press, Berkeley, 1980

20. Dobratz, B.M., LLNL Explosives Handbook, Report UCRL-52997, Lawrence Livermore National Laboratory, March 1981

21. Kohn, B.J., Compilation of Hugonit Equations of State, Report AFWL-TR-69-38, Air Force Weapons Laboratory, 1969

22. Walters, W.P. and Zukas, J.A., Fundamentals of Shaped Charges, Wiley Interscience, New York, 1989

23. Green, A.E. and Naghdi, P.M., A General Theory of an Elastic-Plastic Continuum, Arch. Rat. Mech. Anal., Vol. 2, p. 197ff, 1965

24. Dienes, J. K., On the Analysis of Rotation and Stress Rate in Deforming Bodies, Acta Mechanica, Vol. 32, p.217ff, 1979

25. Herrmann, W., Constitutive Equations of the Dynamic Compaction of Ductile Porous Materials, J. Appl. Phys., Vol. 40, #6, 1969

26. Gupta, Y.M. and Seaman, L, Local Response of Reinforced Concrete to Missile Impact, Report EPRI-NP-217, Electric Power Research Institute, 1979

27. Wright, J.P. and Baron, M.L.,Dynamic Deformation of Materials and Structures Under Explosive Loading, in High Velocity Deformation of Solids, (Eds. Kawata, K. and Shioiri, J.), Springer-Verlag, Berlin, 1979

28. Johnson, G.R. and Cook, W.H., A Constitutive Model and Data for Metals Subjected to Large Strains, High Strain Rates and High Temperatures, Proc. 7th Intl. Symposium on Ballistics, The Hague, The Netherlands, 1983

29. Johnson, G.R. and Cook, W.H., Fracture Characteristics of Three Metals Subjected to Various Strains, Strain Rates, Temperatures and Pressures, Engng. Fract. Mech., Vol. 21, pp. 31-48, 1985

30. Lips, H.R. et al, Dynamic Behavior and Properties of Heavy Metals - Experimental Approach to Separation of Parameters in the Johnson-Cook Model, Proc. 10th Intl. Symp. on Ballistics, Vol. II, San Diego, 1987

31. Johnson, G.R. and Holmquist, T.J., Evaluation of Cylinder-Impact Test Data for Constitutive Model Constants, J. Appl. Phys., Vol. 64, pp. 3901-3910, 1988

32. Rajendran, A.M., Material Constitutive Models, in Material Behavior at High Strain Rates, short course notes, Computational Mechanics Associates, Baltimore, MD, 1993

33. Nicholas, T. and Rajendran, A.M., Material Characterization at High Strain Rates, Chapter 3, High Velocity Impact Dynamics, (Ed. Zukas, J.A.), Wiley Interscience, New York, 1990

34. Zerilli, F.J. and Armstrong, R.W., Dislocation-Mechanics-Based Constitutive Relations for Material Dynamics Calculations, J. Appl. Phys., Vol. 61, pp. 1816-1825, 1987

35. Bodner, S.R. and Partom, Y., Constitutive Equations for Elastic-Viscoplastic Strain-Hardening Materials, J. Appl. Mech., Trans. ASME, Vol. 42, pp. 385-389, 1975

36. Bodner, S.R. and Merzer, A., Viscoplastic Constitutive Equations for Copper With Strain Rate History and Temperature Effects, J. Engng. Mat. Tech., Trans. ASME, Vol. 100, pp. 388-394, 1978

37. Bodner, S.R. and Aboudi, J., Stress Wave Propagation in Rods of Elastic-Viscoplastic Material, Int. J. Solids Struct., Vol. 19, pp. 305-314, 1983

38. Nicholas, T. Rajendran, A.M. and Grove, D.J., Analytical Modeling of Precursor Decay in Strain-Rate Dependent Materials, Int. J. Solids Struct., Vol. 23, pp. 1601-1614, 1987

39. Nicholas, T., Rajendran, A.M. and Gróve, D.J., An Offset Yield Criterion From Precursor Decay Analysis, Acta Mech., Vol. 69, pp. 205-218, 1987

40. Rajendran, A.M. and Grove, D.J., Bodner-Partom Viscoplastic Model in the STEALTH Finite Difference Code, Report AFWAL-TR-86-4089, Wright-Patterson AFB, OH, 1987

41. Steinberg, D.J., Cochran, S.G. and Guinan, M.W., A Constitutive Model for Metals Applicable at High Strain Rates, J. Appl. Phys., Vo. 51, pp. 1498-1504, 1987

42. Steinberg, D.J., Constitutive Model Used in Computer Simulation of Time-Resolved Shock Wave

Data, Int. J. Impact Engng., Vol. 5,#1-4, pp. 603-612, 1987

43. Steinberg, D.J., Equation of State and Strength Properties of Selected Materials, Report UCR-MA-106439, Lawrence Livermore National Laboratory, 1991

44. Follansbee, P.S. and Kocks, U.F., A Constitutive Description of the Deformation of Copper Based on the Use of the Mechanical Threshold Stress as an Internal State Variable, Acta Metall., Vol. 36, p. 81ff, 1988

45. Meyer, L.W., Constitutive Equations at High Strain Rates, in Shock-Wave and High Strain rate Phenomena in Materials, (Eds. Meyers, M.A., Murr, L.E., and Staudhammer, K.P.), Marcel Dekker, New York, 1992

46. Woodward, R.L., Material Failure at High Strain Rates, Chapter 2, High Velocity Impact Dynamics, (Ed. Zukas, J.A.), Wiley Interscience, New York, 1990

47. Poth, A. et al, Experimental and Numerical Investigation of the Ricochetting of Projectiles from Metallic Surfaces, Proc. 6th Intl. Symp. on Ballistics, Orlando, FL, 1981

48. Poth, A. et al, Failure Behavior of an Aluminum Plate Under Impact Loading, Proc. Intl. Conf. on Application of Fracture Mechanics to Materials and Structures, Freiburg, Germany, 1983

49. Ringers, B.E., New Sliding Surface Techniques Enable Lagrangian Code to Handle Deep Target Penetration/Perforation Problems, in Computational Aspects of Penetration Mechanics, (Eds. Chandra, J. and Flaherty, E.), Springer-Verlag, Heidelberg, 1983

50. Bertholf, L.D. et al, Kinetic Energy Projectile Impact on Multi-Layered Targets: Two-Dimensional Stress Wave Calculations, Report SAND76-0375, Sandia National Laboratories, 1976

51. Hancock, J.W. and MacKenzie, A.C., On the Mechanisms of Ductile Failure in High-Strength Steels Subjected to Multi-Axial Stress States, J. Mech. Phys. Solids, Vol. 24, pp. 147-169, 1976

52. Nash, M.A. and Cullis, I.G., Numerical Modeling of Fracture - A Model for Ductile Fracture in Triaxial States of Stress, Proc. 3rd Conf. on Mechanical Properties at High Strain Rates, (Ed. Harding, J.), Institute of Physics Conference Series #70, Institute of Physics, London, 1984

53. Matuska, D.A. and Osborn, J.J., HULL/EPIC Linked Eulerian/Lagrangian Calculations in Three Dimensions, Report ARBRL-CR-00467, U.S. Army Ballistic Research Laboratory, 1981

54. Oscarson, J.H. and Graff, K.F., Report BAT-197A-4-3, Battelle Memorial Institute, 1968

55. Tuler, F.R. and Butcher, B.M., A Criterion for the Time Dependance of Dynamic Fracture, Int. J. Fract. Mech., Vol. 4, pp. 431-437, 1968

56. Mescall, J. and Papirno, R., Spallation in Cylinder-Plate Impact, Exp. Mech., pp. 257-266, July 1974

57. Johnson, G.R. and Stryk, R.A., User Instructions for the EPIC-2 Code, Report AFATL-TR-86-51, Eglin Air Force Base, FL, 1986

58. Curran, D.R., Dynamic Fracture, Chapter 9, Impact Dynamics, Zukas, J.A. et al, pp. 333-366, Wiley Interscience, 1982

59. Rajendran, A.M., Dietenberger, M.A. and Grove, D.J., A Void-Growth Based Failure Model to Describe Spallation, J. Appl. Phys., Vol. 65, pp. 1521-1527, 1989

60. Rajendran, A.M. Grove, D.J. and Dietenberger, M.A., A Dynamic Plasticity Based Failure Model, in Advances in Plasticity, (Eds. Khan, K.S. and Tokuda, M), Pergamon Press, Elmsford, New York, 1989

61. Rajendran, A.M., Dietenberger, M.A. and Grove, D.J., Results From The Recently Developed Dynamic Failure Model, in Shock Compression of Condensed Matter - 1989, (Eds. Schmidt, S.C., Johnson, J.N. and Davison, L.W.), pp. 373-376, Elsevier Science Publishers, London, 1989

62. Grove, D.J., Rajendran, A.M. and Dietenberger, M.A., Numerical Simulation of a Double Flyer Impact Experiment, ibid

63. Breed, B.R., Mader, C.L. and Venable, D., Technique for the Determination of Dynamic-Tensile-Strength Characteristics, J. Appl. Phys., Vol. 38, pp. 3271-3275, July 1967

64. Mader, C.L., Numerical Modeling of Detonation, U. of California Press, Berkeley, 1979

65. Mader, C.L., Neal, T.R. and Dick, R.D. (Eds.), LASL PHERMEX Data, Volume 1, U. of California Press, Berkeley, 1980

66. Nicholas, T., Material Behavior at High Strain Rates, Chapter 8, Impact Dynamics, (Ed. Zukas, J.A.), pp. 277-332, Wiley Interscience, New York, 1982

67. Nicholas, T., Dynamic Tensile Testing of Structural Materials Using a Split-Hopkinson Bar Apparatus, Report AFWAL-TR-80-4053, Air Force Materials Laboratory, Wright-Patterson Air Force Base, 1980

68. Nicholas, T. and Rajendran, A.M., Material Behavior at High Strain Rates, short course notes, Computational Mechanics Associates, Baltimore, MD, 1994

69. Bell, J.F. , Report BRL-CR-184, U.S. Army Ballistic Research Laboratory, 1974

70. Clifton, R.J., Constitutive Models for Plastic Flow at Ultra-High Strain Rates, in Dynamic Constitutive/Failure Models, (Eds. Rajendran, A.M and Nicholas, T.), Report AFWAL-TR-88-4229, Wright-Patterson AFB, 1988

71. Kim, K.S. and Clifton, R.J., Pressure-Shear Impact of 6061-T6 Aluminum, J. Appl. Mech., Trans. ASME, Vol. 47, pp. 11-16, 1980

72. Li, C.H., A Pressure-Shear Experiment for Studying the Dynamic Plastic Response of Metals at Shear Strain Rates of 10^5/s, PhD Dissertation, Brown U., 1982

Chapter 2

Blast on surface structures

A. Barbagelata, M. Primavori

D'Appolonia S.p.A., Via Siena, 20, 16146 Genoa, Italy

ABSTRACT

State of the art techniques for the analysis of a blast wave impinging on surface structures are summarized. The way a blast wave is generated and its interaction with surface structures are described; approximate solutions for some common structures and basic references are provided. Available computer codes for the dynamic analysis of structures subjected to blast waves are reported and commented. The last section of this chapter is devoted to the description of approximate analytical solutions of problems frequently encountered in engineering practice such as overturning of structures due to blast waves and determination of surface temperatures of bodies exposed to fireballs.

INTRODUCTION

In the design and analysis of surface structures to resist the effects of explosions, the principal effects of the explosive output to be considered are blast pressures and primary fragments. Of these two parameters, the blast pressures are usually the governing factor in the determination of the structure's response. However, in some situations, primary fragments may be just as important as the pressures in determining the configuration of the interested structure. This chapter deals primarily with the blast pressures and associated structure loadings produced by explosions. The structure is assumed to be far enough from the explosion so that the blast wave is completely formed and primary fragments have lost their energy.

Three explosive sources which generate blast waves are described in the following: high explosives, fuel-air explosives, and nuclear bombs. A common approach is therefore described for the analysis of surface structures subjected to blast waves, obtained by integrating the basic mechanism of reflection of a plane wave against an infinite wall by means of empirical-experimental considerations. Useful approximate solutions are provided for some common structures. Computational techniques are required when direct experimental tests are not available: a selection of computer programs for the analysis of the interaction of the blast wave with the structure and for the dynamic analysis is provided at the end of this chapter.

EFFECTS OF EXPLOSIONS

The sudden release of energy due to the reaction of explosion, chemical in the case of conventional weapons and nuclear in the case of nuclear bombs, causes a considerable local increase of temperature and pressure. In the case of solid and liquid conventional weapons the explosive material is rapidly converted in a hot, high pressurized gas whose volume, suddenly increased, gives rise to a shock wave propagating at high velocity in the surrounding medium (Henrych [1]). This shock wave, usually referred to as a "blast wave" since it resembles and is accompanied by a strong wind, is characterized by a sudden increase of pressure at the front which gradually decreases with time, as shown in Figure 1. The overpressure represents the excess over the atmospheric pressure; the maximum value (at the shock front) is called the "peak overpressure". As the blast wave travels in the air away from the origin of explosion the overpressure at the front decreases and at some distance behind the shock front the air pressure falls below the ambient pressure, reaching a negative value. At the end of the negative phase, during which the air is sucked in instead of being pushed away from the explosion, the pressure returns to the ambient value.

As the blast wave moves, the mass of the air behind the blast front produces a wind, whose pressure is depending upon the density and the velocity of the air and is known as dynamic pressure. Usually the positive phase of the dynamic pressure is longer than for the overpressure. During the negative phase (t_o^- in Figure 1) the wind blows toward the center of the explosion, caused by a vacuum produced. The peak values of overpressure and of dynamic pressure in the negative phase are smaller than the corresponding of the positive phase. An important parameter in determination of the effects of airblast on structures is the impulse, defined as the area under the pressure-time curve (i and i⁻ in Figure 1).

Explosive materials

High explosives High explosives (such as dynamite, nitro-glycerine, trinitrotoluene known as TNT etc.) are high density materials and, therefore, have a high energy content per unit volume: most of them are solid at room temperature, but they can also be gaseous, liquid or gel in rare cases. These condensed phase explosives explode at a constant detonation velocity, varying between 6.5 and 8 kilometers per second (Baker et al. [2]). The pressure during the detonation process reaches values of the order of one million atmosphere because of the very high initial density while the gaseous products of the explosion still have a high density, very close to that of the solid charge. The effects of solid high explosives like TNT are well known in terms of pressures, impulses and duration more than other liquid explosives. This knowledge is generally extended to other detonating materials by relating the explosive energy of the "effective charge weight" of those materials to that of an equivalent weight of TNT through the following formulation:

$$W_E = \frac{H_{EXP}}{H_{TNT}} W_{EXP}$$ (1)

where W_E and W_{EXP} are respectively the effective and real charge weights of the explosive while H_{EXP} and H_{TNT} correspond to the explosive and TNT heats of detonation. Values of heat of detonation for typical explosive materials are $1.97 \ 10^6$, 2.31 10^6, $2.27 \ 10^6$ ft-lb/lb respectively for TNT, PETN and RDX (belonging to pentolite class). Since other factors may affect the energy equivalence, including material shape (round, square, flat), the number of explosive items, explosive confining (casing), the above formulation is firstly applicable to explosives at the same TNT conditions.

Fuel-air explosives   Fuel-air explosives (FAE) are chemical compounds which use the oxygen of the environment as a primary oxidizing agent. Gaseous fuel-air explosives commonly used in fuel-air weapons are ethylene or propilene oxide and normal propyl-nitrate. With the use of fuel-air explosives no cratering and ejecta phenomena occur (Hartenbaum [3]). The sound speed, the density, the physical size of the products of the detonation, and the energy partition all determine the rate of decay of the peak pressure curve. Following ignition of a flammable vapor cloud a flame propagates through the flammable region at a speed depending on the intensity of turbulence ahead of the flame. Significant overpressure caused by the flame will occur if the flame accelerates to speeds of the order of 100 meters per second or more (Wheatley and Webber [4]). The detonability of liquid fuels is influenced by a variety of factors including vapor pressure, drop size, and molecular structure. Detonability can be improved by adding a sensitizer to the fuel. The fuel is often detonated by one or more detonations after some delay to allow mixing the fuel with air. The heat of combustion of most hydrocarbons is larger than TNT's and this is the reason why fuel-air explosives are very effective (Baker et al. [2]).

Nuclear bombs   The origin of the energy resulting from a nuclear explosion is in the extremely exothermic nuclear reactions within the bomb. In a thermonuclear device these reactions are a combination of fission reactions of heavy metals (uranium and plutonium) and fusion reactions in light materials (hydrogen and tritium) (Messenger and Ash [5]). The destructive action of nuclear weapons is mainly due to blast or shock, as in conventional types of explosives. However there are several basic differences between nuclear and conventional explosions: nuclear bombs can even be millions of times more powerful than the largest conventional weapons and, for the emission of the same amount of energy, a much lower mass is sufficient. Moreover in a nuclear explosion in air only a part of the total nuclear energy arising from the charge is converted into the explosive wave energy (about 50 percent), while the remaining is emitted in the form of light and heat (about 35 percent) and radioactive radiations (about 15 percent). At last, only a small fraction is converted into electromagnetic pulse form (Henrych [1],   Glasstone and Dolan [6], Rudie [7]).

A comparison of overpressure and impulse versus distance for equivalent masses of nuclear, TNT and FAE explosives is shown in Figure 2. It can be noted that for short distance radii the impulse due to FAE explosions envelops those of TNT and nuclear explosions.

The scaling law
A useful property of blast waves is the scaling law which allows the shock or blast wave parameters to be scaled from one reference bomb yield to another, for both chemical or nuclear detonations. Pertinent airblast parameters are ordinarily presented in terms of a reference weapon yield, usually 1 kiloton or 1 megaton. In the international system of units, 1 kt = 4.184 x 10^{12} joules (where 1 US ton = 880 kg). These parameters can then be determined for other yields of interest by scaling from the reference yield values.

If certain assumptions are made regarding the properties of air and shock wave propagation, it can be concluded that a given pressure will occur at a distance from an explosion that is proportional to the cube root of the energy yield. High-explosive and nuclear tests have shown this relation to hold approximately true for yields up into the megaton range. Accordingly, the range or distance, R, at which a particular overpressure or dynamic pressure will occur due to detonation of a weapon of yield W may be found from (Messenger and Ash [5], AFWL [8]):

$$\frac{R}{R_1} = \left[\frac{W}{W_1}\right]^{1/3} \tag{2}$$

where R_1 is the range at which the pressure of interest occurs due to the reference yield, W_1.

Cube root scaling such as the above can also be applied to the airblast time of arrival t_A, positive phase duration t_o and impulse i:

$$\frac{t}{t_1} = \left[\frac{W}{W_1}\right]^{1/3} \quad and \quad \frac{i}{i_1} = \left[\frac{W}{W_1}\right]^{1/3} \tag{3}$$

where t denotes time of arrival or positive phase duration, and i is the impulse of interest; the subscripted terms are the reference yield values. In using the foregoing expressions, it must be understood that both distance and time are being scaled.

Prediction techniques
The most important airblast parameters are: peak overpressure, peak dynamic pressure and pressure-time history. The peak pressures and the pressure time history determine the impulse present in an airblast wave. Other free-field airblast parameters which are

frequently important in specific applications include the shock front velocity and the particle (or wind) velocity behind the shock front.

The pressure rises very sharply at the moving front and falls off toward the interior region of the explosion. The transient pressure in excess of the ambient is defined as the overpressure whose maximum value is the peak overpressure. As the blast wave arrives at a given point, the overpressure suddenly reaches the peak overpressure with a rise time less than a microsecond in the case of nuclear explosions and few milliseconds for conventional explosions. The peak overpressure as a function of weapon yield and distance from the burst point can be found from semiempirical correlations. A rough approximation is presented in Figure 2a which shows the overpressure p_{so} due to an explosion of 1 kiloton yield. Different weapon yields can be accounted for using the scaling law.

The variation of overpressure with time can be expressed as follows (AFWL [8]):

$$p(t) = p_{so}(1-\tau)(ae^{-\alpha\tau} + be^{-\beta\tau})\tag{4}$$

where p_{so} is the peak overpressure; a, b, α, β are a function of peak overpressure; t_o is the duration of the positive phase; τ corresponds to the nondimensional abscissa t/t_o (Figure 1).

Even if the most destructive effects of a blast wave are due to overpressure, in some cases the forces due to dynamic pressure, resulting from the mass air flow behind the shock front, can be of great importance. The decay of dynamic pressure as sum of exponential function of time can be expressed as follows (AFWL [8]):

$$q(t) = q_o(1-\omega)^2(de^{-\delta\omega} + fe^{-\varphi\omega})\tag{5}$$

where q_o is the peak of dynamic pressure; a, b, δ, φ are a function of peak of dynamic pressure; t_u is the duration of the positive phase; ω corresponds to the nondimensional abscissa t/t_u (Figure 1).

LOADS ON STRUCTURES

As the wave front deriving from an unconfined explosion reaches a structure, a portion of the structure or the structure as a whole will be engulfed by the shock pressure. The loads applied to the structures which the blast wave intercepts are depending on the orientation, geometry, and size of the objects the wave encounters. Three main kinds of unconfined explosions are considered (see Figure 3):

- Free air burst load. The blast wave propagates away from the center of the explosion striking the structure without intermediate amplification of the initial shock wave.

- Air burst load. The explosion is located at a distance away from and above the structure so that ground reflections of the initial wave occur before the blast wave reaches the structure.

- Surface burst load. The explosion is located close to or on the ground so that the shock wave is amplified at the point of detonation due to ground reflections.

Blast loadings

Free air burst If the shock wave impinges on a rigid surface oriented at an angle to the direction of propagation of the wave, a reflected pressure is instantly developed on the surface, and the pressure is raised to a value in excess of the incident pressure. The reflected pressure is a function of the pressure in the incident wave and the angle formed between the rigid surface and the plane of the shock front. For a reflector, where flow around an edge or edges occurs, the duration of the reflected pressures is controlled by the size of the reflecting surface. The high reflected pressure seeks relief toward the lower pressure regions, and this tendency is satisfied by the propagation of rarefaction waves from the low- to high-pressure region. These waves, traveling at the velocity of sound in the reflected pressure region, reduce the reflected pressures to the stagnation pressure which is the value that is in equilibrium with the high-velocity air stream associated with the incident pressure wave. When such a relief is not possible, for example when an incident wave strikes an infinite surface, the incident pressure at every point in the wave will be reflected, and these reflected pressures will last for the duration of the wave.

The peak positive reflected pressure is designated p_r, the peak negative reflected pressure p_r^-, and the unit impulse associated with a completely reflected incident wave is i_r for the positive phase and i_r^- for the negative phase. The pressure-time variation for free air burst and for infinite plane reflectors is shown in Figure 4.

When the shock wave impinges on a surface oriented so that a line which describes the path of travel of the wave is normal to the surface, then the point of initial contact is said to sustain the maximum (normal reflected) pressure and impulse. The positive phase pressure and impulse patterns on the structure vary with distance from a maximum at this normal distance to a minimum (incident pressure) where the plane of the structure's surface is perpendicular to the shock front. Under this condition, the instantaneous peak value of the reflected overpressure p_r is given by (Glasstone and Dolan [6]):

$$p_r = 2p + (\gamma + 1)q \tag{6}$$

where p and q refer to incident pressure and dynamic pressure respectively and $\gamma = 1.4$ for air. Introduction of the Rankine-Hugoniot conditions based on the conservation of mass, energy and momentum at the shock front leads to the following expression for the dynamic pressure:

$$q = \frac{p^2}{2\gamma p_a + (\gamma - 1)p} \tag{7}$$

Substitution of the above expression of q in Equation (6) yields:

$$p_r = 2p\frac{7p_a + 4p}{7p_a + p} \tag{8}$$

where p_a is the ambient pressure.

The variation of the pressure and impulse patterns on the surface between the maximum and minimum values is a function of the angle of incidence α. This angle is formed by the line which defines the normal distance between the point of detonation and the structure and the line which defines the path of shock propagation between the center of the explosion and any other point in question.

The effect of the angle of incidence on the peak reflected pressures is shown in Figure 5 which is a plot of the angle of incidence versus the peak reflected pressure coefficient as a function of the peak incident pressure. The peak reflected pressure $p_{r\alpha}$ is obtained by multiplying the peak reflected pressure coefficient C_r by the peak incident pressure p_{so}. The values of the coefficient of reflection with $\alpha = 0$ can be obtained by means of Equation (8).

Air burst Oblique reflection (occurring when blast waves strike a surface at oblique incidence) is classed as either regular or Mach reflection (Courant and Friedrichs [9]). In regular reflection the incident shock, on contact with the wall, is reflected from the wall at a reflection angle different from the incidence. For a given strength of incident shock there is some critical angle of incidence above which regular reflection cannot occur: in these cases the reflected wave, traveling through air heated and compressed, overtakes the incident blast wave creating a third front, called "Mach front". The region where the fusion takes place is called the Mach region; the point at which the two waves fuse is the triple point (Baker et al. [2], Glasstone and Dolan [6]). A representation of an air burst environment is shown in Figure 6. The height of the Mach front increases as the wave propagates away from the center of detonation. A structure is subjected to a plane wave when the height of the triple point exceeds the height of the structure. The structures above the path of the triple point experience the shock from the incident and the reflected wave. For determining the magnitude of the blast loads acting on the surface of a structure, the peak incident blast pressure is computed using the distance and the yield of explosion. Once the incident free air pressure is obtained, then the peak reflected pressure is computed from the diagrams of Figure 5, using the values of the angle of incidence α and the peak incident free air pressure.

Surface burst A charge located on or very near the ground surface is considered to be a surface burst. The initial wave of the explosion is reflected and reinforced by the ground surface to produce a reflected wave. Unlike the air burst, the reflected wave merges with the incident wave at the point of detonation to form a single wave hemispherical in shape. The charge reflection method is used to estimate the magnitude of the blast parameters, assuming that the explosive output of a detonation on the ground is equivalent to that produced by a larger quantity of explosive detonated in free air.

In the case of airblast, the external loads are independent of the structure motion, while, in the case of ground shock, loads and structure motion are dependent from each other and their interaction must be considered. The shock wave front can be considered plane when the structure is located relatively far from the point of detonation: in this case the shock front is assumed to be normal to the ground surface. The forces acting on a structure associated with a plane shock wave are dependent upon the peak values and the pressure variation with time of the incident and of the dynamic pressures.

Loading on above ground structures For above ground closed rectangular structures, the front wave encounters first the front wall (facing the point of detonation), then the sides, the roof and at last the rear wall of the structure. At the moment the incident shock front strikes the wall the pressure is immediately raised from zero to the reflected pressure value. At the same time these surfaces are subjected to drag pressure. The total load on each wall results from the algebraic sum of the overpressure and drag pressure.

Common assumptions made to reduce the problem to reasonable terms include the following (AFM [10]):

a)   the structure is generally considered rectangular in shape,

b)   the incident pressure of interest is in the order of 140 Newtons per square centimeter or less, and

c)   the object being loaded is in the region of the Mach reflection.

Front wall For a rectangular above ground structure at low pressure ranges, the variation of pressure with time on the side facing the detonation is illustrated in Figure 7. At the moment the incident shock front strikes the wall, the pressure is immediately raised from zero to the reflected pressure p_r, which is a function of the incident pressure and the angle of incidence between the shock front and the structure face (Figure 5). The clearing time t_c required to relieve the reflected pressures is represented as:

$$t_c = \frac{3S}{U} \qquad (9)$$

where U is the velocity of the shock front and S is equal to the height of the structure H_s or one-half its width W_s, whichever is smaller. The pressure p acting on the front wall after time t_c is the algebraic sum of the incident pressure p_s and the drag pressure $C_D q$.

$$p = p_s + C_D q \tag{10}$$

The drag coefficient C_D gives the relationship between the dynamic pressure and the total translational pressure in the direction of the wind produced by the dynamic pressure and varies with the Mach number (or with the Reynold's number at low incident pressures) and the relative geometry of the structure.

Roof and side walls As the shock front traverses a structure, a pressure is imparted to the roof slab and side walls equal to the incident pressure at a given time at any specified point reduced by a negative drag pressure. The portion of the surface loaded at a particular time is dependent upon the location of the shock front and the wave lengths (L_w and L_w^-) of the positive and negative pulses.

An equivalent pressure versus time is shown in Figure 8. The pressure builds up linearly from the time t_f when the blast wave reaches the beginning of the element (point f) to time t_d when the blast wave reaches point d. The peak value p_o of the pressure is the sum of the contributions of the equivalent incident and drag pressures.

$$p_o = C_E p_{sob} + C_D q_{ob} \tag{11}$$

where p_{sob} is the peak overpressure occurring at point b and q_{ob} corresponds to the value of $C_E p_{sob}$. Following the peak, the pressure decays linearly to zero at time $t_b + t_{of}$ where t_b is the time at which the blast wave reaches the end of the element (point b) and t_{of} is the fictitious duration of the positive phase. The negative phase of the equivalent uniform loading, if required, may conservatively be taken as that occurring at point b.

The drag coefficient C_D takes values in the range between -0.2 and -0.4 depending on the peak dynamic pressure (AFM [10]). The equivalent load factor C_E is a function of the ratio between the wave length and the length of the structure: it takes values between 0.6 and 0.9.

Rear wall The pressure time history acting on the rear wall of the structure is similar to that shown in Figure 8 for side walls and roof. Here the peak pressure of the pressure-time curve is calculated using the peak pressure that would occur at distance H_s (point e in Figure 8) past the rear edge of the roof slab t_f is replaced by t_b.

COMPUTER CODES FOR STRUCTURAL ANALYSIS AND WAVE STRUCTURE INTERACTION

Two main classes of computer codes are considered in this section:

- computer codes for structural analysis,
- computer codes for wave-structure interaction.

The first class of computer codes can be used for dynamic structural analyses when the loads on the structure are known: for instance they might have been determined by means of the approximate method described in the previous sections. The main features of computer codes for such application include:

- 2-D and 3-D modeling,
- linear and non-linear transient dynamic analyses including large displacements and material behavior.

A list of the most popular general purpose finite element computer codes is presented in Table 1. It is a choice of the authors to avoid a detailed description of the capabilities of the different codes. The information on these codes is generally not homogeneous and this might lead to unfortunate conclusions regarding the suitability of the codes to specific purposes. Names and addresses of the developers are however provided in order to enable the reader to collect additional information on the codes.

The second class of computer codes can be used for the analysis of interaction between the blast wave and the structure. A list of computer codes available for such analysis is provided in Table 2. Most of the codes reported in Table 2 can also be applied to simulate near or contact explosions as well as projectile penetration.

The wave propagation codes were originally developed to solve problems characterized by (Zukas [11]):

- the presence of shock waves (alternatively, steep stress, or velocity gradients),
- localized materials response (i.e., situations where the overall geometric configuration of a structure is of secondary importance compared to the constitution and characteristics of the material in the vicinity of the applied load), and
- loading and response times in the submillisecond regime.

The two methods of solution used in all production codes for wave propagation and impact studies are the finite difference and finite element methods. A common property of both the finite difference and finite element methods is the local separation of the spatial dependence from the time dependence of the dependent variable (i.e., we have a semidiscrete system, algebraic expressions for the spatial behavior of unknowns but still partial derivatives for the time dependence). This permits treatment of space and time grids separately.

Since for many cases the discrete forms of the equations of motion of the finite element method are equivalent to those of the finite difference method, there is no basic mathematical difference between the two methods. Therefore, they should have the same degree of accuracy in numerical computations. The main differences lie not in the methods themselves but in the data management structure of the computer programs that implement them.

Finite element codes have a distinct advantage in treating irregular geometries and variations in mesh size and type. This is because in the finite element method, the equations of motion are formulated through nodal forces for each element and do not depend on the shape of the neighboring mesh. In the finite difference method, equations of motion are expressed directly in terms of the pressure gradients of the neighboring meshes. This is not inherently a problem, but the difference equations must be formulated separately for irregular regions and boundaries. Another major difference occurs in numbering of meshes. In finite difference programs, the regularity of the mesh implicitly establishes the connectivity information. In finite element programs, mesh connectivity is explicitly stored, a feature that facilitates automatic generation of complex mesh systems.

Spatial discretization can be carried out in Eulerian or Lagrangian framework. In the Lagrangian scheme a grid is embedded in the material and distorts with it. In the Eulerian approach the grid is fixed in space and the mass flows through it. Both schemes have advantages and disadvantages for various problem classes. Often, features of both are required in a calculation so the tendency today is toward development of computer codes with both Eulerian and Lagrangian capabilities. There are a number of advantages to a Lagrangian system. Since there are no convective terms for the motion of material through a grid as in the Eulerian case, Lagrangian codes are conceptually straightforward and, in principle, should require fewer computations per cycle. Since the grid deforms with the material, time histories are easily obtained and material interfaces and geometric boundaries are sharply defined. However, in order to account for momentum transfer between colliding bodies, Lagrangian codes incorporate elaborate logic for sliding interfaces.

It should be noted that the wide range of applicability of the wave codes is somehow limited by the difficulty in identifying the material parameters for non linear-elastic behavior, and the computer time needed to reach the solution. Care should be taken in any application to limit model complexity and accurately select the integration time step.

SPECIAL TOPICS

Overturning Stability of a Structure Subjected to a Shock Wave
The evaluation of the overturning stability of a structure subjected to a pressure wave generated by an explosion can be performed by means of a simplified analytical approach.

When the explosion occurs far from the considered structure, the shock wave can be considered plane and its propagation is parallel to the ground. The way the pressure wave acts on the structure depends on various parameters such as size, shape and orientation of the structure and peak overpressure. The analytical model representing the problem is shown in Figure 9.

Assuming that the structure subjected to the pressure wave can be considered as a rigid body, and assuming also that the stabilizing moment can be expressed as linear function of the rotation angle α, it is possible to write the equation of motion of the body subjected to the pressure wave as:

$$\ddot{\alpha} - \frac{Pb}{I\,\overline{\alpha}}\alpha = \frac{M(t)}{I} - \frac{Pb}{I} \tag{12}$$

where $\ddot{\alpha}$ is the second derivative of α with respect to the time; P is the body weight; I is the body rotational inertia with respect to the center of rotation of the structure; $M(t)$ is the applied moment; b and $\overline{\alpha}$ are the geometrical quantities shown in Figure 9. With the additional assumption that the overturning moment can be expressed as:

$$M(t) = M_o e^{-\xi t} \tag{13}$$

Equation (12) can be integrated and the result may be expressed by means of the nondimensional parameters η_1 and η_2:

$$\frac{\alpha}{\overline{\alpha}} = \frac{\eta_1}{\eta_2^2 - 1} x^{\eta_2} + \frac{\eta_1 + \eta_2 - 1}{2(1 - \eta_2)} x + \frac{\eta_1 - \eta_2 - 1}{2(1 + \eta_2)} x^{-1} + 1 \tag{14}$$

where:

$$\lambda^2 = \frac{Pb}{I\,\overline{\alpha}}; \qquad x = e^{-\lambda t} \tag{15}$$

$$\eta_1 = \frac{M_o}{Pb}; \qquad \eta_2 = \frac{\xi}{\lambda} \tag{16}$$

By deriving Equation (13) with respect to the time, the following relation is obtained for the time of zero rotation speed:

$$\eta_1 = \frac{D(\eta_2^2 - 1)}{D(\eta_2 - 1) - 2\eta_2} \tag{17}$$

where:

$$D = \frac{x^2 - 1}{x^2 - x^{\eta_2 + 1}} \tag{18}$$

By substituting the values of η_1, η_2 and x in Equation (14) it is possible to produce a chart for the maximum relative rotation versus the parameters η_1 and η_2. This corresponding diagram is shown in Figure 10. The region of the graph above the curve "Overturning limit" (underlined in Figure 10) corresponds to value of rotation which cause overturning of the structure (Barbagelata and Perrone [12]).

Finite Element Approach The interaction of the blast wave with the structure can be studied directly applying to it the time history of the forces resulting from the explosion. Pressures acting on front, rear, top and bottom faces of the structure can be computed taking into account wave reflection effects, the exposure, and the interaction of the structure surfaces with the pressure wave.

For an accurate finite element analysis of the problem it is necessary to realize a non-linear model able to represent:

- transient pressure loads;
- unilateral contact and friction factors between supports and ground;
- non linear elastic behavior of supports;
- dampers;
- large displacement and rotations.

Non-linearity of the model does not involve the load applied to the structure. The finite element model, shown in Figure 11, includes:

- one mass element with rotational inertia, placed in the center of mass of the structure;
- two non linear spring elements representing the non linear load-displacement characteristic of the supports;
- two damper elements modeling, with an appropriate load-speed law, the energy dissipation of the supports;
- two gap elements with friction factor;
- beam elements.

The large displacements option allows to take into account the increase of the overturning force lever arm, increase due to the structure rotation, and to consider the displacement of the center of rotation in case of sliding.

Thermomechanical Effects of Explosions on Structures
The stresses produced on structures by thermal radiation originated by an explosion may be of great relevance, particularly for FAE and nuclear explosions. A considerable amount of the energy released from a nuclear explosion is emitted in form of heat: a percentage equal to about thirty percent of the total energy is converted into thermal energy. The variation with time of the total energy released from the explosion is shown in Figure 12a.

An analytical solution employing method of superposition of the effects for the problem of a flat plate exposed to a thermal radiation is presented (Barbagelata and Perrone [13]).

The transient thermal radiation is assumed to vary linearly with time, as shown in Figure 12b. The area under the flux-time curve is the thermal impulse I. The thickness of the plate is L and the variation with time t of the thermal radiation can be expressed as:

$$P(t) = \beta k t \tag{19}$$

where k is the thermal conductivity of the material and β is a dimensional coefficient.

The temperature of the exposed surface is given by the following nondimensional expression:

$$\tau = \frac{\theta}{3} + \frac{\theta^2}{2} - \frac{1}{45} + \frac{2}{\pi^4} \sum_{n=1}^{\infty} \frac{e^{-n^2 \pi^2 \theta}}{n^4} \tag{20}$$

where:

$$\tau = \frac{a}{\beta L^3} T; \tag{21}$$

$$\theta = \frac{a}{L^2} t; \tag{22}$$

a is the thermal diffusivity of the material; T is the wall temperature and t the time.

Analyzing the results of Equation (20) it can be observed that the maximum temperature occurs in the decreasing part of the impulse curve; the maximum wall temperature increases as the ratio $\frac{\theta_1}{\theta_2}$ increases. A simplified approximate relation can be obtained:

$$\Delta T_{max} = 0.82 \left(\frac{I P_{max}}{k \rho c} \right)^{1/2} \tag{23}$$

where ρ is the density of material; c is the specific heat; P_{max} is the maximum heat flux absorbed by the plate and I is the total heat energy absorbed. It can be observed that the value of ΔT_{max} does not depend on the thickness of the plate.

The maximum wall temperature occurs at a time shift from the beginning of the thermal transient and conservatively it can be assumed that the mean temperature does not change. Deformations on the surfaces can be computed as:

$$\varepsilon = \alpha \Delta T \tag{24}$$

where α is the coefficient of thermal expansion of the material.   The corresponding stresses in the plane of the plate, assuming elastic behavior of the material, are:

$$\sigma = \frac{\varepsilon}{1-\upsilon} E \tag{25}$$

where E is the Young modulus and ν the Poisson ratio.   These stresses are usually very high but they do not cause collapses of the structure due to the fact that only limited portions of the material are involved.

The global effects in the structure can be evaluated assuming a uniform temperature increment in the wall:

$$\Delta T = \frac{I}{L\rho c} \tag{26}$$

The analysis of the simultaneous effects induced by both the blast wave and the thermal pulse may be required for structures exposed to nuclear explosions in the following cases:

- the temperature reached by the material changes either its strength or stiffness or both,
- stresses induced by temperature gradients increase significantly those induced by the blast wave and may induce damage in the material,
- structural deformation is a concern.

All computer codes listed in Table 1 can be used to perform a thermo-mechanical analysis of a structure.

CONCLUSIONS

A brief description of state of the art techniques for the analysis of the effects of a blast wave on surface structures has been given in this chapter.   The way blast waves are generated by high explosives, fuel-air explosives, and nuclear bombs has also been described.   If the explosion is sufficiently far from the target, the blast wave propagation and the wave-structure interaction are independent on the explosion characteristics.   The most commonly used approach to determine the loads induced on the structure by the blast wave are based on the generalization of the interaction of a plane shock front with an infinite plate.   The loads on the structure surfaces can be computed analytically by means of few simple formulas summarized in this chapter.   If a more accurate computation of the loads is required, the use of the so-called wave codes is recommended.   A list of structural codes and wave codes to be used for dynamic analyses and wave-structure interaction analyses has been provided.   Since in almost all applications the time required to develop the model and to run the analysis is important, analytical solutions of simplified problems should be used, whenever available, to gain a better understanding of the level of criticality of the structure and, therefore, on the level of accuracy needed.

ACKNOWLEDGMENT

For helpful criticism and advice on the manuscript, the authors would like to express their thanks to their friends and colleagues Gian Maria Manfredini and Fabrizio Lagasco.

References

1.   Henrych, J., The Dynamics of Explosion and Its Use, Elsevier, Amsterdam Oxford New York, 1979.

2.   Baker, W. E., Cox P.A., Westine P. S., Kulesz J. J. and Strehlow R. A., Explosion Hazards and Evaluation, Elsevier, Amsterdam Oxford New York, 1983.

3.   Hartenbaum, B. A., Nuclear Airblast Simulation Using Fuel Air Explosives, Defense Nuclear Agency, Washington DOC 203895, 1976.

4.   Glasstone, S. and Dolan P. J., The Effects of Nuclear Weapons, United States Department of Defense and Energy Research and Development Administration, Washington, 1977.

5.   Messenger, G. C. and Ash, M. S., The Effects of Radiation on Electronic Systems, Van Nostrand Reinhold, New York, 1992.

6.   Wheatley, C. J. and Webber, D. M., Aspects of the Dispersion of Denser than Air Vapours Relevant to Gas Cloud Explosions, Commission of the European Communities, Nuclear Science and Technology, 1984.

7.   Rudie, N. J., Principles and Techniques of Radiation Hardening, Western Periodicals Company, North Hollywood (CA), 1986.

8.   Air Force Weapons Laboratory (AFWL), The Air force Manual for Design and Analysis of Hardened Structures, 1974.

9.   Courant R. and Friedrichs K. O., Supersonic Flow and Shock Waves, Interscience Publishers Inc., New York, 1948.

10.   Department of the Air Force Manual, Structures to Resist the Effects of Accidental Explosions, AFM 88-22, Departments of the Army, the Navy, and the Air Force, Washington, 1969.

11.   Zukas, J. A., High Velocity Impact Dynamics, John Wiley & Sons Inc., New York, 1990.

12.   Barbagelata, A. and Perrone, M., A Nonlinear Finite Element Approach to the Assessment of Global Stability of a Military Vehicle Under Shock Wave Impulse, pp. 497 to 507, Proceedings of the 2nd Int. Conf. on Structures Under Shock and Impact, Portsmouth, United Kingdom, 1992.   Computational Mechanics Publications and Thomas Telford, London, 1992.

13.   Barbagelata, A. and Perrone, M., Analisi degli Effetti Termomeccanici di un' Esplosione su una Lastra Piana, X Congresso Nazionale AIMETA, Pisa, 1990.

Table 1. General purpose finite element programs

PROGRAM NAME	DEVELOPER
ABAQUS	Hibbit, Karlsson and Sorensen Inc. - 1080 Main Street - Pawtucket, RI 02860-4847 - U.S.A.
ADINA	Adina R. & D. Inc. - 71 Elton Avenue - Watertown, MA 02172 - U.S.A.
ALGOR	Algor Inc. - 150 Beta Drive - Pittsburgh, PA 15238-2932 - U.S.A.
ANSYS	Swanson Analysis System Inc. - P.O. Box 65 - Houston, PA 15342-0065 - U.S.A.
ASKA	IKO Software Service GmbH - Albstadtweg 10 - D7000 Stuttgart 80 - Germany
CASTOR	CETIM - 52 Avenue Félix Louat - 60300 Senlis - France
COSMOS	Structural Research and Analysis Corporation - 2951 28th St. Suite 1000 - Santa Monica, CA 90405 - U.S.A.
DAPSYS	D'Appolonia S.p.A. - Via Siena, 20 - 16146 Genova - Italy
MARC	Marc Analysis Research Corporation - 260, Sheridan Avenue, Suite 309 - Palo Alto, CA 94306 - U.S.A.
NASTRAN	The MacNeal-Schwendler Corporation - 815, Colorado Boulevard - Los Angeles, CA 90041 - U.S.A.
NISA	Engineering Mechanics Research Corporation (EMRC) - Troy, MI 48099 - U.S.A.

Table 2. Wave codes

PROGRAM NAME	DEVELOPER
AUTODYN	Century Dynamics Inc. - 7700 Edgewater Drive, Suite 626 - Oakland, CA 94621 - U.S.A.
CTH	Sandia National Laboratories - P.O. Box 5800 - Albuquerque, NM 87185 - U.S.A.
DEFEL	Dyna East Corporation - 3201 Arch Street - Philadelphia, PA 19104-2855 - U.S.A.
DYNA 2D/3D	Lawrence Livermore National Laboratory - P.O. Box 808 - Livermore, CA 94550 - U.S.A.
DYSMAS	Industrieanlagen-Betriebsgesellschaft (IABG) mbH - Ottobrunn - West Germany
EPIC-2/3	Honeywell Inc., Defense Systems Division - 7225 Northland Drive - Brooklyn Park, MN 55428 - U.S.A.
HELP	Systems Science and Software - La Jolla, CA - U.S.A.
HEMP/3D	Lawrence Livermore National Laboratory - Livermore, CA 94550 - U.S.A.
HULL	Orlando Technology Inc. - P.O. Box 855 - Shalimar, FL 32579 - U.S.A.
METRIC	Systems, Sciences and Software Inc. - La Jolla, CA - U.S.A.
PISCES2/3DELK	PISCES International Company - 2700 Merced Street - San Leandro, CA 94577 - U.S.A.
STEALTH	Science Applications Inc. - San Leandro, CA - U.S.A.
SOIL	Computer Code Consultants Inc. - 820 Windcrest - Carlsbad, CA 92009 - U.S.A.
TOODY-4	Sandia National Laboratories - P.O. Box 5800 - Albuquerque, NM 87185 - U.S.A.
ZEUS	Computational Mechanics Consultants Inc. - P.O. Box 11314 - Baltimore, MD 21239 - U.S.A.

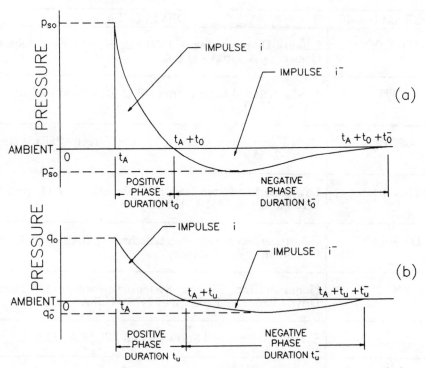

Figure 1: Static overpressure (a) and dynamic (b) pressures-time variation.

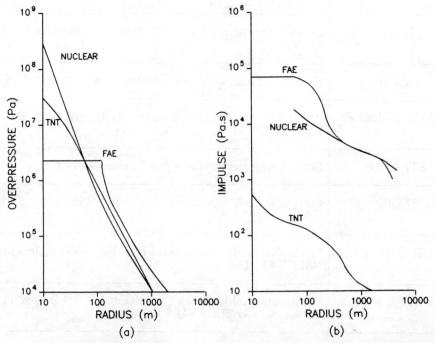

Figure 2: Static overpressure (a) and impulse (b) for 1 kiloton nuclear, TNT, and FAE explosions.

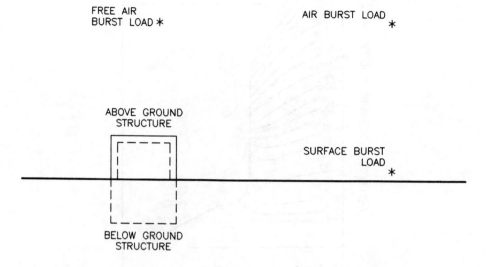

Figure 3: Blast loading categories.

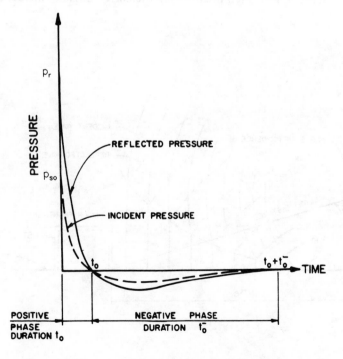

Figure 4: Pressure-time variation for free air burst.

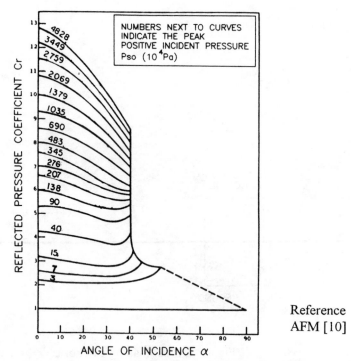

Reference
AFM [10]

Figure 5: Reflected pressure coefficient vs. angle of incidence.

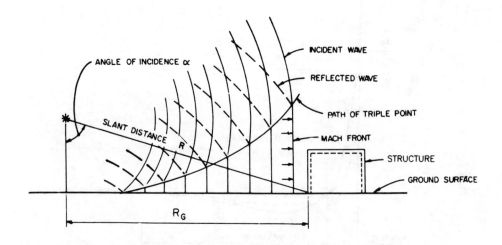

Figure 6: Air burst blast environment.

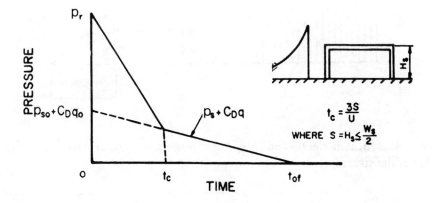

Figure 7: Front wall loading.

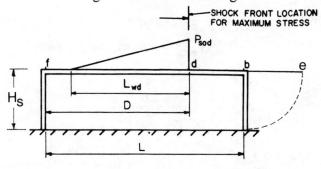

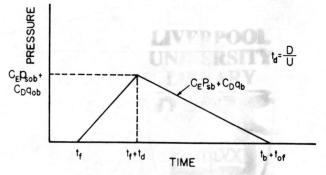

Figure 8: Roof and side wall loading.

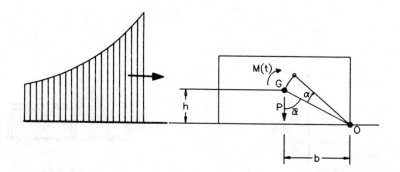

Figure 9: Analytical model (G and O represent respectively the centers of mass and rotation of the structure).

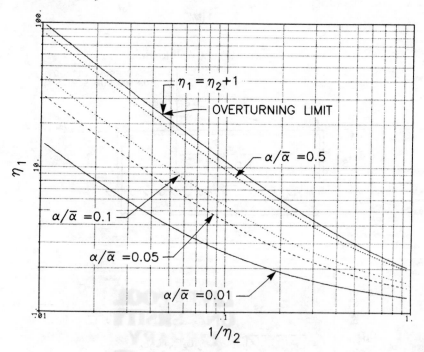

Figure 10: Charts for relative rotation.

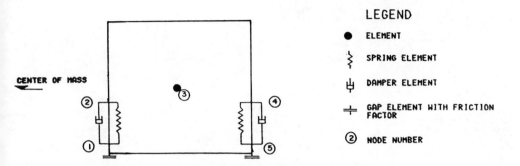

Figure 11: Finite element model.

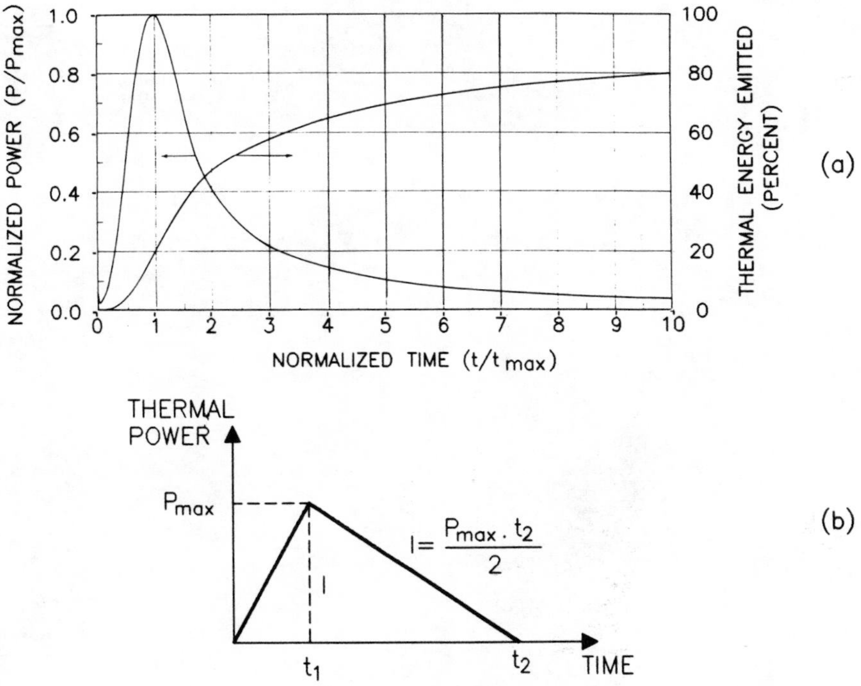

(a)

(b)

Figure 12: Transient thermal radiation.

Chapter 3

Low velocity perforation of metal plates

N. Jones

Impact Research Centre, Department of Mechanical Engineering, The University of Liverpool, P.O. Box 147, Liverpool L69 3BX, UK

ABSTRACT

This chapter examines the perforation of metal plates struck normally by relatively large 'rigid' masses travelling with low impact velocities which cause large global displacements prior to perforation. Recent studies on the development of empirical equations which are suitable for design purposes, the influence of nearby supports on the perforation energy and the laws of geometrically similar scaling are discussed.

NOTATION

c	elastic wave speed
d	diameter of the impact face of a blunt striker
k	kH is critical transverse shear displacement, $0 < k \leq 1$
r	radial coordinate measured from centre of a circular plate
E	energy
E_c	dynamic energy required to produce a visible crack on the distal side of a plate
E_p	dynamic energy required for perforation
E_{pp}	predicted value of E_p according to the geometrically similar scaling laws from the smallest scale model
E_s	static energy required for perforation
F	impact force
G	mass of striker
H	plate thickness
R	radius of the supporting boundary of a circular plate
S	unsupported span, as defined in Figure 1
V_o	initial impact velocity
V_p	impact velocity required for perforation
β	geometrical scale factor ($\beta \geq 1$)
λ	ratio of plug mass to the projectile mass
ξ	r/R
ρ	density of plate material
σ_y	uniaxial yield stress of the plate material

σ_u ultimate tensile stress of the plate material
Δ transverse displacement of striker

INTRODUCTION

This chapter focuses principally on the behaviour of a ductile metal plate which is perforated by a rigid striker travelling with a relatively low impact velocity. This structural geometry is used for safety calculations and hazard assessments in several industries in which plates or panels may be struck by dropped objects or by masses propelled by the gases escaping after an explosion. Many articles have been published on the perforation of plates due to high velocity impacts [1-6, etc.], but the associated theoretical analyses disregard usually the influence of global displacements which give rise to membrane forces within a plate. However, membrane forces are an important energy absorbing mechanism for low velocity impacts whether or not perforation occurs, and, therefore, should be retained in the governing equations [7, 8].

The dynamic perforation of ductile metal plates is a complex phenomenon and numerical and theoretical studies are still continuing in this area [e.g., 6, 9]. These and many other recent studies focus largely on the high or moderately high velocity regime and explore the detailed local behaviour around the striker and through the plate thickness, but, as noted above, often ignore any global response. However, recent studies [8, 10-16] have examined the dynamic plastic response and perforation of thin plates struck by relatively heavy strikers travelling with low velocities.

Various empirical equations have been developed for the perforation of metal plates because of the complexity of the phenomenon. The validity of these equations is well documented for the higher impact velocity events [1, 17], but the accuracy of these equations for heavy masses travelling with low velocities has not been well established. In some cases, the empirical equations were obtained from experiments on plates which suffer different phenomena to those associated with low velocity impacts. Consequently, different empirical equations have been published for the low velocity impact range [13, 14, 16, 18]. These equations have different ranges of validity, sometimes unknown, which are related to the material properties of the plate and the structural details. This is a confusing and unsatisfactory situation and is due to the difficulty of modelling the complex dynamic behaviour and developing a satisfactory theoretical analysis.

It is the objective of this chapter to provide an overview of some recent studies on the low velocity impact perforation of thin plates. The next section examines the different failure modes which are possible in thin metal plates subjected to large impact loads. Some comments then follow on empirical equations to predict perforation, the influence of impacts near to supporting boundaries and on the accuracy of the geometrically similar scaling laws for the impact perforation of plates. This chapter is then completed with a discussion section and a conclusion.

FAILURE

A ductile metal plate when subjected to an impact load may fail in one of several modes. The external dynamic load may cause the development of membrane forces and large permanent transverse displacements without any rupture of the material. This behaviour is sometimes known as a mode I response and may occur in any ductile structure and is a type of failure if the magnitude of the permanent displacements is restricted in a particular design. Rigid-plastic methods of analysis [7,19,20] have been found to give good agreement with experiments which have been performed on mild steel and aluminium alloy circular and rectangular plates subjected to blast loadings [21,22].

For sufficiently large dynamic loads, a local tensile tearing may occur at the boundary of a plate or on the surface of a plate underneath a striker. This is known as a mode II tearing failure and occurs when the rupture strain of the material is exceeded. Little work has been done on this failure mode for plates, but it has been studied for beams subjected to local impact loads [23, 24] and it has been observed in some experimental tests on circular plates subjected to blast loads [25].

It has been found at still higher impact loads, that a transverse shear, or mode III, failure may occur around the impact zone, or at the supports. This type of failure has been studied using rigid-plastic methods of analysis for circular plates [26,27] and has been observed in experimental tests on circular plates subjected to blast loads [25]. A mode III transverse shear failure occurs when the total transverse shear displacement at a particular location equals the structural thickness, or some smaller critical thickness associated with failure [7], as discussed in Reference [28] for the impact loading of beams. For example, if the critical transverse shear displacement equals kH, where $0 < k \leq 1$ and H is the plate thickness, then it is shown in Reference [29] that the theoretical analysis in Reference [26] predicts a critical velocity for perforation

$$V_p = \{k\pi\sigma_y H^2 d(1+\lambda)/G\}^{1/2} , \tag{1}$$

where σ_y is the uniaxial yield stress of the plate material, d is the diameter of a striker having a mass G and λ is the ratio of the plug mass to the projectile mass. Equation (1) is identical to the theoretical predictions of Recht and Ipson [30] when k = 0.5.

It may be shown, when using the equations in §6.4.2 of Reference [7] for a long rigid, perfectly plastic beam impacted by a rigid mass at the mid-span, that a transverse shear failure would occur at small transverse displacements of the beam when the impact velocities are large as in the foregoing analysis in which they were neglected altogether. The equations show, for small impact velocities, that the associated transverse displacements of the beam would be significant at severance. Similar characteristics are expected to occur in a circular plate impacted by a mass. This explains why global displacements and the influence of membrane forces are neglected in analyses for the perfora-

tion of plates at high impact velocities, but are important for low
impact velocities.

It is evident, therefore, that the energy absorbing and failure
characteristics of plates perforated by masses with high and low impact
velocities, are quite different, and require distinct theoretical treat-
ments. As noted earlier, emphasis has been focused largely on the high
velocity case because of military interest, but some recent attention
has been given to the low velocity case which is relevant to many
civilian problems.

EMPIRICAL EQUATIONS

The perforation of a plate by a striker, or projectile, as shown in
Figure 1, is a highly complex process, as noted in the previous sections. Thus, no comprehensive theoretical models are available which
incorporate all of the phenomena and which are capable of predicting
accurately all of the features of an impact perforation event. However,
designers in engineering practice require guidance on the perforation of
plates which has led to the development of empirical equations [1].

Two fairly simple formulae that have been used widely in design are
the SRI (Stanford Research Institute) and BRL (Ballistics Research
Laboratory) equations [17, 31, 32] which are given by equations (2) and
(3) in Table 1, respectively. The SRI equation is based on the experimental data from more than six hundred tests covering a range of
projectile and target sizes. The parameter range for the BRL equation
is less clear. Neilson [32] has proposed equation (4) in Table 1 for
the perforation of steel plates by a long penetrator, while Jowett [17]
has derived equations (5) and (6) in Table 1 for the perforation of mild
steel plates by short missiles. The recent formulation given by
equation (7) in Table 1 was developed in Reference [18] for the perforation of steel plates impacted by a blunt or flat-ended projectile, as
shown in Figure 1. The form of this equation was suggested by the observation in References [5, 8, 14, 15, 33] that, for low impact velocities,
there is a significant dishing displacement in the plates as well as a
plugging failure at the impact site. The first term in equation (7) is
an estimate of the energy required to shear a plug out of a plate, while
the second term represents the energy absorbed by a global structural
response including bending and stretching effects.

The coefficients in equation (7) were selected to fit the new
experimental data presented in Reference [18] for steel plates with S/d
= 40 and $0.39 \leq H/d \leq 1.57$ and impact velocities up to 12.2 m/s. It is
evident from Figure 2 that, apart from equation (7), only the simple one
term BRL equation (3) gives reasonable agreement with the experimental
results. However, it is shown in Reference [18] that equation (7) also
gives good agreement with the experimental data on the perforation of
steel plates reported by Langseth and Larsen [8] (S/d = 14.5, $0.11 \leq H/d$
≤ 0.27, $8.5 \leq V_o \leq 39.5$ m/s), Corran et al. [15] (S/d = 19.2, $0.1 \leq H/d$
≤ 0.51, $40 \leq V_o \leq 180$ m/s, approximately) and Neilson [32] ($4.0 \leq S/d \leq$
34.9, $0.025 \leq H/d \leq 0.397$, $9.5 \leq V_o \leq 102$ m/s). Again, apart from

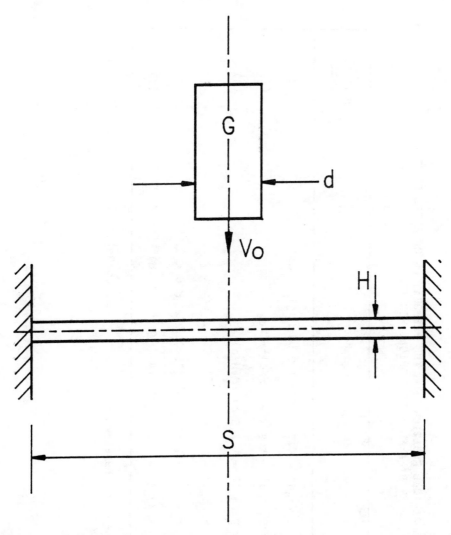

Figure 1:   A plate with a uniform thickness H which is supported across a span S (diameter, if circular) and struck by a rigid mass G having a diameter d and travelling with an initial velocity V_o.

equation (7), only the BRL equation (3) gives reasonable agreement with all of this experimental data.   Nevertheless, the experimental results do, in many cases, lie outside the range of validity of the various empirical equations, as noted in Reference [18]. No limitations appear to be associated with the BRL equation (3) [17,18] and the range of validity of equation (7), for low impact velocities, is unknown.

It is evident that none of the right hand sides of the empirical equations (2) to (7) in Table 1 is related to the perforation velocity V_p.   However, the experimental results in Figure 3 reveal that it is an important factor which leads to a range of dimensionless impact

TABLE 1 - EMPIRICAL EQUATIONS FOR THE PERFORATION OF STEEL PLATES
BY A FLAT CYLINDRICAL MISSILE

Reference	Empirical equation[+]	Equation Number
SRI [17,31,32]	$E_p/\sigma_u d^3 = 4.15(H/d)^2 + 0.097(S/d)(H/d)$	(2)
BRL [17,31,32]	$E_p/\sigma_u d^3 = 1.4 \times 10^9 (H/d)^{1.5}/\sigma_u$	(3)
Neilson [32]	$E_p/\sigma_u d^3 = 1.4(S/d)^{0.6}(H/d)^{1.7}$	(4)
AEA (Jowett) [17]	$E_p/\sigma_u d^3 = 6.0(H/d)^{1.74}, \; 0.1 \leqq H/d < 0.25$	(5)
AEA (Jowett) [17]	$E_p/\sigma_u d^3 = 1.73(H/d)^{0.84}, \; 0.25 < H/d < 0.65$	(6)
Wen and Jones [18]	$E_p/\sigma_u d^3 = 2(\sigma_y/\sigma_u)\{(\pi/4)(H/d)^2 + (S/d)^{0.21}(H/d)^{1.47}\}$	(7)

+: restrictions on the use of these equations are given in References [17] and [18] and in the original references

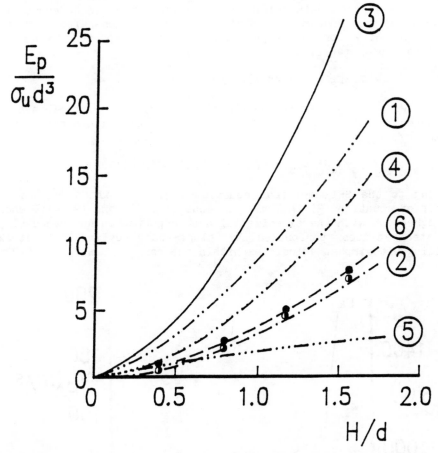

Figure 2:   Comparison of various empirical formulae with the test data
in Reference [18] for the perforation of mild steel plates struck at the
centre (S = 203.2 mm, d = 5.08 mm,
G = 5.6 kg, 2 ≤ H ≤ 8 mm).

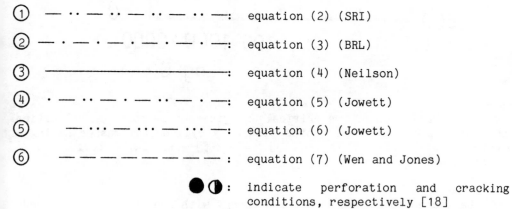

① —‥—‥—‥—‥—: equation (2) (SRI)

② —·—·—·—·—: equation (3) (BRL)

③ —————————: equation (4) (Neilson)

④ ·—‥—·—‥—·—: equation (5) (Jowett)

⑤ —‥‥—‥‥—‥‥—: equation (6) (Jowett)

⑥ ———————— : equation (7) (Wen and Jones)

●◑: indicate   perforation   and   cracking
conditions, respectively [18]

energies for a given value of H/d, as shown further in Reference [13]. The ratio of the dynamic to static perforation energies versus the critical velocity for the experimental results reported in Reference [13] obeys approximately the linear relation

$$E_p/E_s = 1 + 48.78(V_p/c) ,\qquad(8)$$

where $c = \sqrt{E/\rho}$ = 5135 m/s is the elastic wave speed for steel when E and ρ are taken as 207 GPa and 7850 kg/m³, respectively. The static perforation energy, E_s, in equation (8) is taken as

$$E_s/\sigma_u d^3 = 2.075(H/d)^2 + 0.0485(S/d)(H/d)\qquad(9)$$

according to the SRI empirical equation [13, 34], which is, in fact, one-half of equation (2) in Table 1. Equations (8) and (9) give encouraging agreement with the experimental data reported in Reference [13] on mild steel plates perforated by blunt-ended cylindrical strikers travelling with impact velocities within the range $7.7 \leq V_p \leq 118.9$ m/s.

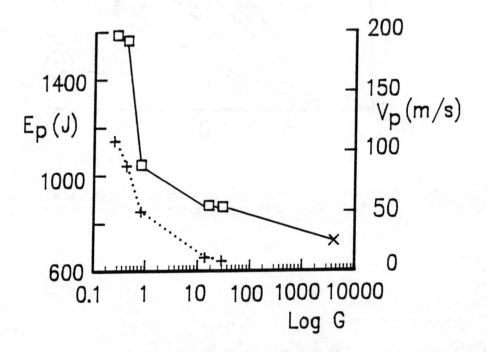

Figure 3: Variation of the critical energies and perforation velocities for 6 mm thick mild steel plates struck by cylindrical masses having a blunt end with d = 17.85 mm [13]. ——□——: E_p, ...+.....: V_p, X: static perforation

The increase in the perforation energy with impact velocity according to equation (8) is most probably due to the strengthening influence of material strain rate sensitivity, as discussed further in Reference

[13]. However, a maximum dynamic perforation energy is reached at around 100 m/s in several experimental studies reported in Reference [13], the actual value of which depends on the material and geometrical characteristics of a particular plate perforation problem. A decrease in the dynamic perforation energy then occurs for further increases in the impact velocity and decreases in the striker mass which is due to a change in the failure mode related to thermo-plastic instability (shear banding).

Some experimental results are reported in Reference [14] for aluminium alloy plates with $0.25 \leq H/d \leq 0.79$ and $5.33 \leq S/d \leq 80$ and perforated by blunt cylindrical strikers with impact velocities up to 11.55 m/s. A procedure similar to that outlined in Reference [18] was used to obtained the empirical equation

$$E_p/\sigma_u d^3 = (\sigma_y/\sigma_u)\left[(\pi/4)(H/d)^2 + 0.1(S/d)^{0.6}(H/d)^{1.7}\right] \qquad (10)$$

for the perforation of aluminium alloy plates.

IMPACTS NEAR SUPPORTS

Experimental tests were conducted in Reference [14] on mild steel plates which were struck at several positions along a radial line from the centre up to the supports, while two impact positions were examined for the aluminium alloy specimens at the centre of the plates and at a location close to the supports. It is evident from Figure 4 that the impact energies required for both cracking and perforation decrease with an increase of the parameter ξ, where $\xi = r/R$ and r is the radial distance from the centre of a plate having a radius R. In other words, the energy absorbing capacity of a plate is smaller when it is struck by a mass near to the supports. It was also observed that the drop in energy absorbing capacity between $\xi = 0$ and $\xi \cong 1$ for the 2 mm and 4 mm thick mild steel plates was more pronounced than that for the 6 mm and 8 mm thick plates. In the former case (2 mm and 4 mm plates), the energy required for perforation and cracking was reduced by approximately 30-40%, while in the latter case (6 mm and 8 mm plates) the maximum drop in energy absorbing capacity was less than 20%, approximately. The situation is somewhat similar for the aluminium alloy plates but, in some cases, the perforation energy at $\xi \cong 1$ is only about one-half of the perforation energy at $\xi = 0$. This behaviour is due to two factors. One is that, for plates struck at the centre, membrane forces play a greater role in absorbing the impact energy in the thinner plates. The other factor is that strikes near to a support cause relatively small transverse displacements of a plate and, therefore, most of the impact energy is absorbed through bending and shear effects and not membrane forces regardless of the plate thickness.

The empirical equations (7) and (10), which were obtained in Reference [18] for impacts at the mid-span, were modified in Reference [14] by replacing S by $S(1-\xi)$ to give the respective empirical equations

$$E_p/\sigma_u d^3 = (2\sigma_y/\sigma_u)\left[(\pi/4)(H/d)^2 + \{S(1-\xi)/d\}^{0.21}(H/d)^{1.47}\right] \qquad (11)$$

and

$$E_p/\sigma_u d^3 = (\sigma_y/\sigma_u)\left[(\pi/4)(H/d)^2 + 0.1\{S(1-\xi)/d\}^{0 \cdot 6}(H/d)^{1 \cdot 7}\right] \qquad (12)$$

for mild steel and aluminium alloy plates struck by blunt cylindrical masses at any dimensionless radial position, ξ. A comparison is made in Figure 4 between the predictions of equation (11) and some of the experimental results from Reference [14] on 2 mm thick mild steel plates.

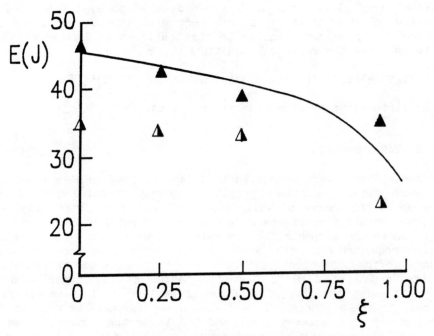

Figure 4: Variation of E_p with E_c with dimensionless impact position (ξ) for the 2 mm thick mild steel plates in Reference [14] with H/d = 0.39. Solid and half symbols indicate perforation (E_p) and cracking (E_c) conditions, respectively. ————: equation (11).

SIMILITUDE

The testing of small-scale models is indispensable for complex structural systems, which are difficult to analyse theoretically and numerically, or to study experimentally. The dynamic response of underground structures, impact of nuclear fuel capsules, missile impact of nuclear power installations, and the collision protection of ships, illustrate several areas which have been studied with the aid of small-scale models.

Dynamic tests are conducted on a small-scale model in order to obtain the response characteristics of a geometrically similar full-scale prototype, which is the actual system of interest. This procedure is known as scaling, modelling or similitude, and is governed by certain principles. Apart from the obvious purpose of relating the behaviour of a model to that of a prototype, these principles also predict various dimensionless combinations of the governing variables which are valuable for the planning of experimental investigations and the choice of numerical calculations.

The standard methods of dimensionless analysis [7] show how the parameters of a structural problem must be varied in order to satisfy the requirements for geometrically similar scaling of a small-scale model and a full-scale prototype. However, it is impossible to scale all the effects simultaneously, and the Buckingham π-theorem and other dimensional methods are incapable of assessing the potential importance of distortion which is produced when all the variables of a problem are not varied according to the laws of geometrically similar scaling. Nevertheless, the experimental structural impact study by Duffey [35] demonstrated that the laws of geometrically similar scaling are satisfied, although it was shown in Reference [36] that his conclusions require some qualification.

Duffey, Cheresh and Sutherland [37] have used the Buckingham π-theorem to generate 16 dimensionless parameters which should be satisfied for the strict scaling of structures loaded dynamically by a punch. The authors also conducted experimental tests on full-scale and one-half scale fully clamped mild steel and stainless steel circular plates subjected to a central impact loading by a solid circular punch. It transpired that the full-scale prototype plate, without any foam backing, suffered permanent transverse displacements which are about 10% larger than those expected from the one-half scale tests when using the laws of geometrically similar scaling. Similarly, the external energy required to cause the failure of a full-scale prototype is smaller than expected.

Booth, Collier and Miles [38] reported a series of drop tests on one-quarter scale to full-scale thin plated mild steel and stainless steel structures. The specimens were geometrically similar, as far as possible, and subjected to the same impact velocity which is required for elementary scaling. Therefore, the permanently deformed shapes of the specimens should be geometrically similar, but they show significant departures from similitude. In fact, the authors found that the post-impact deformations might be as much as 2.5 times greater in a full-scale prototype than would have been expected from the extrapolated results obtained on a one-quarter scale model.

A series of experimental tests on the dynamic cutting of mild steel plate was conducted on a drop hammer rig and reported in References [36] and [39]. The geometric scale factor β is taken as the ratio of the largest (full-scale) plate thickness to the plate thickness of a small-scale model. It is evident from Reference [36] that the principles of elementary scaling predict from the experimental results, with β = 1 that 2.27 times more energy would be absorbed than was actually absorbed during the experiments on the thickest plate with β = 4.03. In other words, the thickest plates absorb only 44 per cent of the energy which would have been expected from small-scale tests on plates having one-quarter of the thickness.

It is interesting to note that the above observations are in broad agreement with several other dynamic and static studies discussed in Reference [36] and that there is quantitative agreement between the significant departures from the principles of geometrically similar scaling observed in the two quite different experimental programmes reported in References [36] and [38].

Some test results are reported in Reference [28] for aluminium alloy and mild steel double-shear specimens subjected to large dynamic loads which produce extensive inelastic behaviour. It was found that the transverse displacements at the initiation of cracking almost obeyed the elementary geometrically similar scaling law. However, the critical displacement for complete transverse severance, or failure, did not, since the fracture displacement, as a proportion of the beam thickness, was less for the thickest specimens. In addition, the results for the small-scale models, when scaled up using the elementary scaling laws, overpredicted the actual energy absorbed in the experimental tests on the full-scale prototypes.

Atkins [40] employed an energy balance approach to examine the scaling laws for combined plastic flow and fracture. The total kinetic energy is equated to the plastic work consumed in the material through structural deformations and the energy required for creating new fracture surfaces. The plastic and fracture energies scale as the cube and square of the scale factor (β), respectively, as noted in Reference [41]. In general, a practical static or dynamic problem involves fracture as well as plastic flow. Thus, the total energy absorbed would scale as β^x, $2 \leq x \leq 3$. Atkins [42] further discussed his theory with respect to the experimental results obtained for the double-shear specimens reported in Reference [28]. It is likely that the scaling laws are problem-dependent with the geometrically similar scaling laws being satisfied in some cases but not in others. The dynamic response of some structures passes through several phases from elastic through elastic-plastic and large plastic strains to rupture with different scaling laws governing the principal phenomena during each phase.

An experimental investigation is reported in Reference [12] into the scaling laws for fully clamped circular plates struck by blunt projectiles travelling at relatively low impact velocities. The test model plates were made from either mild steel (strain rate sensitive) or aluminium alloy (strain rate insensitive). The test specimens and impact conditions were geometrically scaled as far as possible and scale factors of four for the mild steel plates and approximately five for the aluminium alloy specimens were examined. The external kinetic energies ranged from relatively low values giving small plastic deformations in the plates up to high values which produced cracking or perforation.

The experimental results for the transverse deformations, interfacial forces (see Figure 5) response times and perforation energies (see Figure 6) obey the geometrically similar scaling laws within a

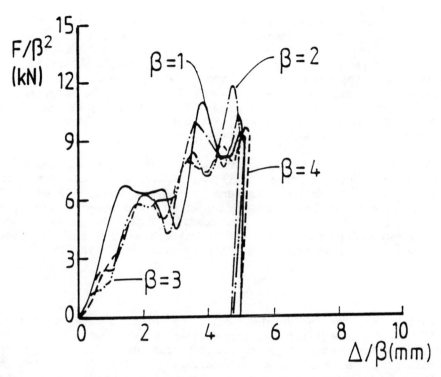

Figure 5:   Scaling comparisons for impact force (F) versus transverse displacement at mid-span (Δ) of the mild steel plates struck at the mid-span in Reference [12].

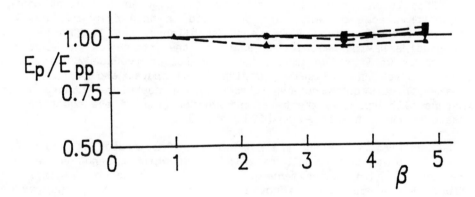

Figure 6: Ratio of the perforation energy (E_p) and the corresponding predicted value (E_{pp}) versus β for geometrically scaled impact tests in Reference [12].   Aluminium alloy specimens with H/d = 0.25. E_{pp} for ▲, ●, ■ and ▼ are based on the experimental results for 2(β = 1), 4.76(β = 2.38), 6.35(β = 3.175) and 9.53 mm (β = 4.765) thick plates, respectively.   ———: E_p/E_{pp} = 1 for geometrically similar scaling.

range of accuracy which is expected for dynamic tests of the type reported here. It is observed that the influence of the mechanical properties of the materials causes a slight deviation from the geometrically similar scaling laws for the aluminium alloy which is less pronounced for the permanent transverse deformations of the mild steel specimens. It also appears that the lack of geometrically similar scaling of material strain rate sensitive effects [7] is not important for the perforation of plates at least within the range of the experimental parameters studied in Reference [12].

The laws of geometrically similar scaling were also examined in Reference [13] but for mild steel plates having a scale range of two struck by blunt cylindrical strikers travelling up to 119 m/s. These experimental results lend further support to the observations reported in Reference [12]. However, it is possible that the energy absorbing mechanism may change for higher impact velocities and, thereby, cause a violation of the elementary geometrically similar scaling laws.

DISCUSSION

This chapter has focused on the perforation of thin metal plates struck normally by blunt cylindrical projectiles travelling with low impact velocities which cause large global deformations. The plate response contrasts with that associated with ballistic perforation at high impact velocities which is dominated by compression of the plate material underneath a blunt striker and transverse shear effects (including possible shear banding) in the vicinity of the striker boundary without any significant global deformations. Despite the restriction of low velocities, the perforation problem is still an important practical one, but a difficult one, for which most progress has been made using empirical equations, some of which are presented in this chapter. However, these empirical equations have been established usually from experimental studies having restricted ranges of the geometrical and material parameters as well as the impact test conditions (boundary conditions, striker location and obliquity, striker nose shape, etc.). Thus, the various empirical equations are valid only for restricted ranges of the parameters although the extent is unknown for some of the equations, as discussed further in References [17] and [18].

The empirical equations are not suitable for predicting perforation in an actual plate which would fail in a mode which was not encountered in the corresponding experimental tests used to develop the empirical equation. For example, it is noted in Reference [13] that the perforation energies from several studies on steel plates increase from static perforation energies with increase in impact velocity up to about 100 m/s, but then decrease for still higher impact velocities. This marked change in behaviour is due likely to the phenomenon of shear banding in the highly distorted zone near the boundary of the striker and requires further study to establish any limitations required on the empirical equations.

Most of the low velocity studies have examined the perforation of plates using blunt-ended cylindrical projectiles. However, Ohte et al. [43] conducted tests on steel plates with cylindrical strikers having several nose shapes and impact velocities between 25 and 180 m/s. They found that the conical strikers required smaller penetration energies than the blunt cylindrical projectiles, while the hemispherically nosed projectiles did not perforate for an impact energy which would have caused a blunt-ended projectile to penetrate a plate. Further tests were reported in Reference [44] in order to assess the importance of different steels for the projectiles and target plates.

Corran et al. [15] have performed tests on 1.3 mm thick mild steel targets struck by d = 12.5 mm diameter cylindrical projectiles having various nose radii from d/2 (hemispherical) to infinity (flat-ended). It was observed that the perforation energy increased about threefold as the nose radius decreased from ∞ to d, approximately. However, the perforation energy decreased quite sharply for even smaller nose radii. The peak value of the perforation energy is associated with a change in the perforation mode from one of shearing around the edge of a projectile at large nose radii to tensile failure at small nose radii.

Corbett and Reid [16] have reported the results of experimental tests on simply supported steel plates struck by hemispherically-tipped and flat-ended cylindrical indenters. The strikers weighed only 72 g so that the critical velocities lay within the range of 35 to 300 m/s. It was observed that the empirical equations were valid over a greater range of impact velocities for the hemispherically-tipped strikers than for the flat-ended ones. For the hemispherical case, perforation energies according to the Neilson and SRI equations in Table 1 were larger than the experimental values, while the BRL and AEA equations in Table 1 were smaller. The strikers were too light and lay outside the range of validity for the empirical equations for the blunt-ended indenters except for the AEA equations in Table 1 which gave good agreement with the critical energies.

Some recent experimental studies [45] on the perforation of 2.5 mm thick steel plates with impact velocities up to 15.3 m/s have shown that the critical energy for a conical striker (90° included angle) is about 75 per cent of the critical energy required for perforation by a cylindrical striker with a blunt end. Other recent studies [46] on 2 mm thick steel plates with impact velocities up to 7.5 m/s have obtained the critical energies for perforation with conical, blunt cylindrical and hemispherically nosed strikers.

CONCLUSIONS

This chapter focuses on the perforation of metal plates struck normally by relatively large 'rigid' masses travelling with low impact velocities which cause the development of large global displacements and membrane forces in a plate. Recent studies on the development of empirical equations, the influence of adjacent supports on the perforation energy and similitude are discussed.

The impact energy required to perforate a plate is significantly smaller for projectile strikes near to a support than for strikes at the mid-span.   It appears from the study reported herein that the laws of geometrically similar scaling are satisfied for the perforation of steel and aluminium alloy plates struck by blunt cylindrical projectiles.

It is evident that further studies on several aspects of the low velocity perforation of thin plates are required but that some empirical equations are available which are suitable for design purposes provided they are used within the associated range of validity.

ACKNOWLEDGMENTS

The author wishes to take this opportunity to thank the Impact Research Centre in the Department of Mechanical Engineering at the University of Liverpool and, in particular, Mrs. M. White, for her assistance with the preparation of this manuscript.

REFERENCES

1.   Backman, M. E. and Goldsmith, W.   The Mechanics of Penetration of Projectiles into Targets, Int. J. Engng. Sci., Vol. 16, pp. 1-99, 1978.

2.   Bai, Y. L. and Johnson, W.   Plugging: Physical Understanding and Energy Absorption, Metals Technology, Vol. 9, pp. 182-190, 1982.

3.   Zaid, A. I. O. and Travis, F. W.   An Examination of the Effect of Target thickness in the Perforation of Mild Steel Plate by a Flat-Ended Projectile, Int. J. Mech. Sci., Vol. 16, pp. 373-383, 1974.

4.   Woodward, R. L.   The Interrelation of Failure Modes Observed in the Penetration of Metallic Targets, Int. J. Impact Engng., Vol. 2, No. 2, pp. 121-129, 1984.

5.   Liss, J. and Goldsmith, W.   Plate Perforation Phenomena Due to Normal Impact by Blunt Cylinders, Int. J. Impact Engng., Vol. 2. No. 1, pp. 37-64, 1984.

6.   Anderson, C. E. Jr., and Bodner, S. R.   Ballistic Impact: The Status of Analytical and Numerical Modelling, Int. J. Impact Engng., Vol. 7, No. 1, pp. 9-35, 1988.

7.   Jones, N., Structural Impact, Cambridge University Press, Cambridge, U.K., 1989.

8.   Langseth, M. and Larsen, P. K.   Dropped Objects' Plugging Capacity of Steel Plates: An Experimental Investigation, Int. J. Impact Engng., Vol. 9, No. 3, pp. 289-316, 1990.

9.   Batra, R. R. C. and Jayachandran, R.   Effect of Constitutive Models on Steady State Axisymmetric Deformations of Thermoelastic-Viscoplastic Targets, Int. J. Impact Engng., Vol. 12, No. 2, pp. 209-226, 1992.

10. Langseth, M. and Larsen, P. K.   'The Behaviour of Square Steel Plates Subjected to a Circular Blunt Ended Load', Int. J. Impact Engng., Vol. 12, No. 4, pp. 617-638, 1992.

11. Langseth, M. and Larsen, P. K. Dropped Objects Plugging Capacity of Aluminium Alloy Plates, Int. J. Impact Engng., Vol. 15, No. 3, pp. 225-241, 1994.

12. Wen, H-M. and Jones, N. Experimental Investigation of the Scaling Laws for Metal Plates Struck by large Masses, Int. J. Impact Engng., Vol. 13, No. 3, pp. 485-505, 1993.

13. Jones, N. and Kim, S-B. An Experimental Study on the Large Ductile Deformations and Perforation of Mild Steel Plates Struck by a Mass, University of Liverpool, Impact Research Centre Report No. IRC/103/93, 1993.

14. Wen, H-M. and Jones, N. Experimental Investigation into the Dynamic Plastic Response and Perforation of a Clamped Circular Plate Struck Transversely by a Mass, proc. I.Mech.E. (In Press).

15. Corran, R. S. J., Shadbolt, P. J. and Ruiz, C. Impact Loading of Plates - An Experimental Investigation, Int. J. Impact Engng., Vol. 1, No. 1, pp. 3-22, 1983.

16. Corbett, G. G. and Reid, S. R. Quasi-Static and Dynamic Local Loading of Monolithic Simply Supported Steel Plate, Int. J. Impact Engng., Vol. 13, No. 3, pp. 423-441, 1993.

17. Jowett, J. The Effects of Missile Impact on Thin Metal Structures, Report UKAEA SRD R378, April 1986.

18. Wen, H-M. and Jones, N. Semi-Empirical Equations for the Perforation of Plates Struck by a Mass. Structures Under Shock and Impact II (Ed. P. S. Bulson), pp. 369-380, Computational Mechanics Publications, Southampton, and Thomas Telford, London, 1992.

19. Jones, N. A Theoretical Study of the Dynamic Plastic Behaviour of Beams and Plates with Finite-Deflections, Int. J. Solids Struct., Vol. 7, pp. 1007-1029, 1971.

20. Jones, N. Recent Studies on the Dynamic Plastic Behaviour of Structures, Applied Mechanics Reviews, Vol. 42, No. 4, pp. 95-115, 1989.

21. Jones, N., Uran, T. O. and Tekin, S. A. The Dynamic Plastic Behaviour of Fully Clamped Rectangular Plates, Int. J. Solids and Struct., Vol. 6, pp. 1499-1512, 1970

22. Nurick, G. N. and Martin, J. B. Deformation of Thin Plates Subjected to Impulsive Loading - A Review. Part I: Theoretical Considerations and Part II: Experimental Studies, Int. J. Impact Engng., Vol. 8, No. 2, pp. 159-186, 1989.

23. Liu, J. H. and Jones, N. Experimental Investigation of Clamped Beams Struck Transversely by a Mass, Int. J. Impact Engng., Vol. 6. No. 4, pp. 303-335, 1987.

24. Jones, N. On the Dynamic Inelastic Failure of Beams. Chapter 5, Structural Failure (Eds. T. Wierzbicki and N. Jones), pp. 133-159, John Wiley, New York, 1989.

25. Teeling-Smith, R. G. and Nurick, G. N. The Deformation and Tearing of Thin Circular Plates Subjected to Impulsive Loads, Int. J. Impact Engng., Vol. 11, No. 1, pp. 77-91, 1991.

26. Jones, N. and de Oliveira, J. G. Dynamic Plastic Response of Circular Plates with Transverse Shear and Rotatory Inertia, J. Appl. Mech., Vol. 47, pp. 27-34, 1980.

27. Li, Q. M. and Jones, N. Blast Loading of Fully Clamped Circular Plates with Transverse Shear Effects, Int. J. Solids and Structures (In Press).

28. Jouri, W. S. and Jones, N.   The Impact Behaviour of Aluminium Alloy and Mild Steel Double-Shear Specimens, Int. J. Mech. Sci., Vol. 30, pp. 153-172, 1988.

29. Jones, N. Some Comments on the Dynamic Plastic Behaviour of Structures (Ed. Z. Zhemin and D. Jing), pp. 49-71. Int. Symp. on Intense Dynamic Loading and Its Effects, Beijing, China, 1986. Science Press, Beijing, China and Pergamon Press, Oxford, 1988.

30. Recht, R. F. and Ipson, T. W. Ballistic Perforation Dynamics, J. Applied Mechanics, Vol. 30, pp. 384-390, 1963.

31. Gwaltney, R. C.   Missile Generation and Protection in Light Water-Cooled Power Reactor Plants, Report ORNL-NSTC-22, Oak Ridge National Lab., Tenn., U.S.A., September 1968.

32. Neilson, A. J. Empirical Equations for the Perforation of Mild Steel Plates, Int. J. Impact Engng., Vol. 3, No. 2, pp. 137-142, 1985.

33. Palomby, C. and Stronge, W. J. Blunt Missile Perforation of Thin Plates and shells by Discing, Int. J. Impact Engng., Vol. 7, No. 1, pp. 85-100, 1988.

34. Zabel, N.R. Containment of Fragments from a Runaway Reactor, Stanford Res. Lab. Tech. Report No. 1, SRIA-1, California, U.S.A., 1958.

35. Duffey, T. A.   Scaling laws for Fuel Capsules Subjected to Blast, Impact and Thermal Loading, SAE Paper No. 719107. pp. 775-786. Proceedings Intersociety Energy Conversion Engineering Conference, 1971.

36. Jones, N. Scaling of Inelastic Structures Loaded Dynamically. Structural Impact and Crashworthiness (Ed. G. A. O. Davies), Vol. 1, Keynote Lectures, pp. 45-74, Elsevier Applied Science Publishers, London, 1984.

37. Duffey, T. A., Cheresh, M. C. and Sutherland, S. H. Experimental Verification of Scaling Laws for Punch-Impact-Loaded Structures, Int. J. Impact Engng., Vol. 2, No. 1, pp. 103-117, 1984.

38. Booth, E., Collier, D. and Miles, J. Impact Scalability of Plated Steel Structures. Chapter 6, Structural Crashworthiness (Ed. N. Jones and T. Wierzbicki), pp. 136-174, Butterworths Publishers, London, 1983.

39. Jones, N. and Jouri, W. S. A Study of Plate Tearing for Ship Collision and Grounding Damage, Journal of Ship Research, Vol. 31, pp. 253-268, 1987.

40. Atkins, A. G. Scaling in Combined Plastic Flow and Fracture, Int. J. Mech. Sci., Vol. 30, pp. 173-191, 1988.

41. Jones, N. Structural Aspects of Ship Collisions. Chapter 11, Structural Crash-worthiness (Ed. N. Jones and T. Wierzbicki), pp. 308-337, Butterworths Publishers, London, 1983.

42. Atkins, A. G. Note on Scaling in Rigid-Plastic Fracture Mechanics, Int. J. Mech. Sci., Vol. 32, No. 6, pp. 547-548, 1990.

43. Ohte, S., Yoshizawa, H., Chiba, N. and Shida, S. Impact Strength of Steel Plates Struck by Projectiles, Bull. J.S.M.E., Vol. 25, No. 206, pp. 1226-1231, 1982.

44. Yoshizawa, H., Ohte, S., Kashima, Y., Chiba, N. and Shida, S. Impact Strength of Steel Plates Struck by projectiles, Bull. J.S.M.E., Vol. 27, No. 226, pp. 639-644, 1984.

45. Birch, R. S. and Jones, N. Hazard Study on Projectile Impact, Impact Research Centre Report (Confidential), Department of Mechanical Engineering, The University of Liverpool, 1994.

46. Jones, N., Birch, R. S. and Wadi, R.A-S. Influence of Projectile Nose Shape on the Ballistic Perforation of Metal Plates, Impact Research Centre Report, Department of Mechanical Engineering (Confidential), The University of Liverpool, 1993.

Chapter 4

The dynamic response of ceramics to shock wave loading

Z. Rosenberg

RAFAEL, P.O. Box 2250, Haifa, Israel

ABSTRACT

The renewed interest in strong ceramics, over the past decade, resulted in a wealth of both experimental and theoretical works concerning their dynamic response to impulsive loading. The article reviews the new techniques which were developed to study the dynamic behaviour of ceramics to shock loading focusing on the plate impact experiment and time-resolved stress and particle velocity measurements. We shall also discuss the modelling of failure criteria as implemented in numerical simulations of processes such as long rod penetration into thick ceramic tiles.

INTRODUCTION

The response of brittle solids to impulsive loading has recently become a subject of great interest through the need to understand the dynamic response of rocks - relevant to the search for new energy sources, and ceramics - for the development of new armour systems and aircraft engine blades.

The earliest comprehensive study on ceramic armour was initiated in the late 1960's by Mark Wilkins and his colleagues at Lawrence Livermore National Laboratory. In a series of reports [1]-[4] they summarized their experimental and theoretical effort to develop an optimal lightweight armour design for helicopters, against small arm bullets. They identified a number of ceramics that performed efficiently against 0.3" armour piercing (AP) bullets, when backed by a layer of a light, and rather soft, material like fiberglass. These ceramics, which included alumina, boron and silicon carbides and beryllium-oxide, were studied extensively in a series of plate impact experiments by Gust, Royce and their co-workers [5]-[7]. Hugoniot curves and their elastic limits (HEL) were determined for these ceramics in the 0-30 GPa pressure range. The HEL of the ceramic specimen was shown, by Wilkins and his colleagues, to play an important role in determining the ballistic efficiency of ceramics. Thus, the work of the LLNL researchers combined one dimensional shock wave studies with 2D code simulations (the HEMP code), in order to understand the axially symmetric terminal ballistic experiment, in which an armour piercing bullet interacts with a hard ceramic tile. Besides the HEL of the ceramic it was found that its spall strength is of major importance for ballistic efficiency. Code simulations showed that increasing the spall strength from 0.3 to 0.5 GPa will improve the tile's resistance to penetration considerably, through the longer time it stays intact, exerting its high compressive strength on the projectile.

The renewed interest in ceramics, as possible candidates for a heavy armour system, has prompted a wealth of experimental and theoretical research programs in many national laboratories throughout the world. New experimental techniques were developed in order to understand and quantify the dynamic response of these strong ceramics to impulsive loading, focusing on their failure as brittle materials. The data from these experiments is used to calibrate failure criteria of the loaded material in numerical simulations and, specifically, to account for its degraded elastic properties after failure.

The purpose of the present paper is to review these experimental techniques and summarize most of their findings which are currently used in code simulations of the complex interaction of modern projectiles with ceramic tiles.

THE RELEVANT DYNAMIC PROPERTIES

The pioneering work of M. Wilkins and his colleagues, together with later works by Mescall and Tracy [8] Chartagnac [9], Shockey *et al* [10], and Forrestal and Longcope [11], have highlighted the important physical processes which take place during the penetration of a projectile through a thick ceramic tile. Fig. (1) shows a cross section of such a tile (taken from [10]) which has been impacted by a long-rod penetrator made of a tungsten alloy. The tile was embedded in a heavy steel confinement and the impact velocity was low enough to prevent penetration of the tile by the rod. As pointed out by the authors of [10] the damage pattern, revealed in this experiment, clearly demonstrates the complex issue of failure in brittle materials subjected to impulsive loading. One can identify a series of cracks (cone, lateral and radial), which are formed far ahead of the penetrator, as well as an appreciable amount of comminuted material just ahead of it. This comminuted material has very different properties from those of the intact ceramic and must therefore be characterized in order to account for the tile's resistance to penetration. Throughout the rest of this review we assume that the reader is familiar with the basic terminology of shock wave physics. A short account is given in the Appendix for those who are not acquainted with these terms.

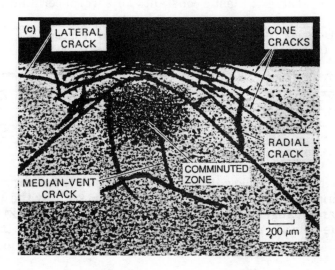

Fig. 1: A cross section of a thick ceramic tile after impact of long rod (from [10]).

The important physical parameters which are needed in order to describe this complex interaction are: (i) the Hugoniot curve of the ceramic and its elastic limit (HEL) which marks the onset of failure under one-dimensional strain loading; (ii) the spall strength of the material which is its dynamic tensile strength under these conditions; (iii) the high pressure shear strength of the intact ceramic and (iv) the shear strength of the comminuted material. The first three parameters determine the failure envelope of the intact material while the fourth accounts for the resistance to penetration of the comminuted zone ahead of the penetrator.

The following sections describe the experimental techniques which are used to quantify these parameters and list some of the relevant data which have been accumulated during the past decade concerning these issues.

HUGONIOT CURVES AND THEIR ELASTIC LIMITS

Although most tough ceramics have relatively low densities (2.5-4.5 g/cc) their dynamic compression curves - the Hugoniot curves - are quite steep owing to their very high sound speeds (9-12 km/s). Table 1 summarizes their relevant elastic constants for some of the more interesting ceramics, together with their HEL values (taken from [12]-[14]),

Table 1: Elastic constants for several ceramics.

Material	Density (g/cc)	Bulk modulus (GPa)	Shear modulus (GPa)	Poisson's ratio	HEL (GPa)
B_4C	2.5	230	199	0.165	16.5
Al_2O_3	3.4 - 4.0	150 - 250	88 - 160	0.22 - 0.25	6.0 - 21.0
SiC	3.2	213	187	0.16	15.5
TiB_2	4.5	215	237	0.1	5.0 - 13.0
AlN	3.25	201	127	0.238	9.0

When shock-loaded below their respective HEL's these ceramics exhibit an elastic loading-unloading behaviour, as measured both by the VISAR (see ([12] and [13]) and manganin stress transducers (see [14]). Fig. (2) (from [13]) shows a particle velocity profile of a low amplitude shock wave traversing a specimen of SiC. Note the almost square shape of the pulse both upon loading and unloading.

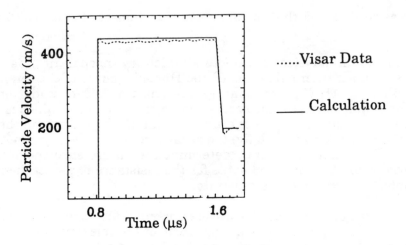

Fig. 2: An elastic loading-unloading wave in a SiC (from [13]).

When shocked to stresses above the HEL (up to about 1.5 times the HEL) most ceramics show a complex wave structure with an elastic jump followed by a more gradual ramping signal. Fig. (3) shows VISAR and manganin traces of waves travelling though Lucalox (polycrystalline alumina [12]) and a commercial alumina (AD85 manufactured by Coors, see [14]). Both measurements clearly show this ramp wave above the HEL. An interesting feature in the release part of these profiles is the relatively large elastic drop in stress or particle velocity, which is a clear indication that the shocked material retains some of its strength even at these high shock stresses.

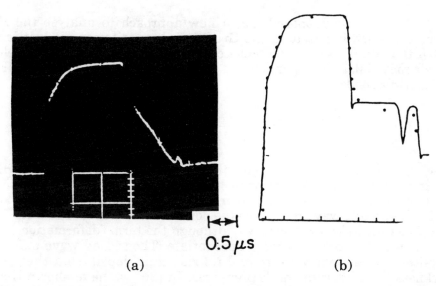

(a)   (b)

0.5 μs

Fig. 3: Stress and particle velocity profiles in a) AD85 with manganin [14] and b) Lucalox using VISAR [12].

The HEL values which were recorded for various ceramics range between 6.0-20 GPa (see Table 1) and are much higher than those obtained for metals (which are in the 0.5-2.5 GPa range). It has been shown that metals and their alloys obey the well-known relation between their HELs (yielding under uniaxial strain loading) and their compressive strength under uniaxial stress conditons (Yo):

$$HEL = \frac{1-v}{1-2v} \cdot Yo \qquad (1)$$

where v is Poisson's ratio. This relation is obtained by applying Tresca or von-Mises yield criteria to the elastic relation between the principal stresses in the shocked specimen:

$$\sigma_Y = \frac{v}{1-v} \cdot \sigma_X \qquad (2)$$

where σ_X and σ_Y are longitudinal and lateral stresses, respectively. Ceramics, on the other hand do not "obey" relation (1) and their measured HEL's were found to be higher by 50-100% than those predicted by Eq. (1). This discrepancy was attributed by most workers to either strain-rate or pressure dependence of their yield strengths. Special experiments were designed to find the cause of this discrepancy either by varying the loading rate - using ramp wave generators [15], or by using long rod techniques to load the ceramic at high strain rates under uniaxial stress loading [16]. These works have shown that the strain rate effect on the strength of ceramics is relatively small, leaving the pressure effect as the only source of strengthening. Thus, numerical modelling of the constitutive relations of ceramics has to rely on "pressure hardening" as was done in [17] and [18], for example.

Recently we proposed [19] a new approach to analyse the HEL of ceramics, which is based on the Griffith failure criterion for brittle solids, rather than the Tresca or von Mises criteria, which account for the yielding of ductile materials. This approach led to a new relation between the HEL and Yo for brittle solids:

$$HEL = \frac{1-v}{(1-2v)^2} \cdot Y_O \qquad (3)$$

which is in better agreement with the experimental values for the HELs of ceramics. When shocked to much higher stresses (above about 1.5 HEL) ceramics show a second wave which follows the ramping increase above the HEL. This second wave resembles the double wave structure in metals and is therefore called the "plastic wave" although the term "deformation wave" or "failure wave" is probably more appropriate. The second wave moves at an amplitude dependent velocity and its rise time depends on the specimen thickness for moderate shock pressures. In [20] we have shown that non-steady state features are obtained for specimens of various thicknesses. These features include phenomena which are common in metals namely, the decay of elastic precursor amplitude and the increasing rise time of the "plastic" wave with specimen thickness. Figure 4 shows three manganin stress gauge records for different thicknesses of alumina specimens with a clear decay of the elastic precursor (up to thicknesses of 10-15 mm) and increasing rise time of the "plastic wave" for these propagation distances.

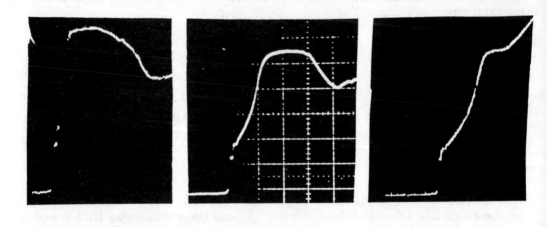

Fig. 4: Precursor decay and non-steady "plastic" waves in alumina specimens - 5, 10, and 15 mm thick. 0.5 μs/div (from [20]).

The only exception for this ramping behaviour of the deformation wave is that of boron carbide. This material shows a distinct two wave structure in which the elastic wave is rather noisy as recorded by both manganin gauges [21] and free surface velocity measurements with VISAR [22]. Figure [5] below shows two typical signals of stress waves in B4C from these works and, as one can see, the elastic and deformation waves are well separated while the elastic waves carry very unusual perturbations. According to Kipp and Grady [22] these are indications of heterogenous yielding processes within the material which cause large particle velocity and stress variations from point to point in the specimen.

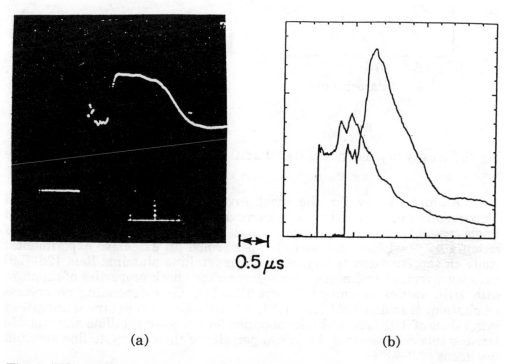

0.5 μs

(a)   (b)

Fig. 5: "Elasto-plastic" waves in B4C: a) manganin gauge record from [21], b) VISAR records from [22].

The relatively large separation between the two waves in boron carbide is a clear indication for an elasto-hydrostatic behaviour of this material, caused by a loss of shear strength at high pressures, as will be discussed later.

One of the most important parameters which affect the HEL of ceramics is their initial porosity. It turns out that the HEL of a given ceramic is strongly dependent on its porosity as shown by both Longy and Cagnoux [23] for Alumina, and by Brar et al [24] for B4C. (see Fig. 6.).

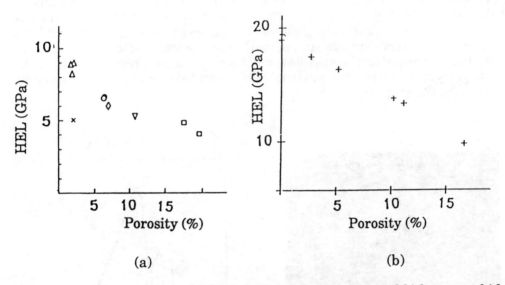

Fig. 6: Porosity dependence of HEL for: a) alumina [23], and b) boron carbide [24].

Alumina is by far the most investigated ceramic material and consequently much work has been carried out with both polycrystalline and single crystal (sapphire) specimens. Much of this work has been summarized recently by Mashimo [25] who also performed an extensive experimental study on sapphire and two types of polycrystalline alumina. Refs. [26]-[29] have used various techniques to determine the shock properties of sapphire with HEL values ranging between 12.0-21.0 GPa depending on crystal orientation. Mashimo [25] finds HEL ≈ 17GPa for his sapphire as compared with values of 6-13 GPa which he obtained for his polycrystalline alumina. He ascribes this difference to the initial porosity of the polycrystalline alumina specimens (2.3-4.3%).

DYNAMIC TENSILE (SPALL) STRENGTH

A typical spall experiment consists of impacting a thin flyer plate on an instrumented specimen disc, about twice as thick. Release waves, from the back surfaces of both impactor and target, meet at about mid-section of the specimen and a tensile wave is generated. If the amplitude of this wave is larger than the dynamic tensile strength of the specimen a spall is created - a new free surface within the bulk of the specimen. Recording either the free surface velocity with VISAR or its stress-time history with a manganin gauge (backed by a softer material) enables one to determine the spall strength through the amplitude of the oscillations in the recorded signals. These measurements have been performed on many metals (see [30]-[32]) resulting in spall strengths of 1.0-5.0 GPa, in accordance with their ultimate tensile strengths.

Since ceramics have extremely low tensile strengths it is not surprising that the measured values of spall strength for polycrystalline ceramics fall in the 0.3-0.6 GPa range (see [12]-[14], [33]-[36]. In [14] we reported that the spall strength of a commercial alumina (AD85, manufactured by Coors) is dependent on the amplitude of the shock wave which loads it prior to the tensile loading. As long as the preloading was smaller than about 1/2 HEL the spall strength was constant (at 0.3 GPa) while for stresses between (0.5-1)xHEL it decreased linearly, reaching zero at a shock level near the HEL. This finding clearly suggests that damage is induced, at least at this type of ceramic, even when shocked below the HEL, and that the HEL should be related to the onset of a total comminution of the specimen. As we shall see below similar conclusions were reached by Luoro and Meyers [36] and Yeshurun *et al* [37], using soft recovery techniques. Moreover, Clifton *et al* [38] have measured free surface velocity profiles of alumina specimens which were subjected to a well defined tensile stress of about 50μs duration, after being shock compressed to only 0.8 GPa. Their profiles indicate that some damage has been induced in the specimen during the loading part. These very small values of spall strength make it extremely difficult to analyse the microstructure of recovered specimens since in most cases they are totally comminuted. This is in sharp contrast with the vast amount of work which has been carried out using recovered metallic specimens over the years (see [39], [40] for example).

PROBING THE STATE OF SHOCKED CERAMICS

Two approaches have been taken in order to overcome this difficulty. In the first, new experimental loading methods are used to probe the state of the ceramic specimen by reloading it with a well defined stress pulse, immediately after the first loading. These are termed the double loading techniques. The second approach is to use novel designs of flyer plates to minimise the damage in the specimen by tensile release waves, in order to recover intact portions of the shock loaded specimen and to analyse it under the microscope. We shall briefly describe these techniques in the next chapter.

Double impact experiments
There are two versions of the double impact experiment; both of them consist of sending a second shock immediately after the passage of the first one, as shown in Fig. 7 below:

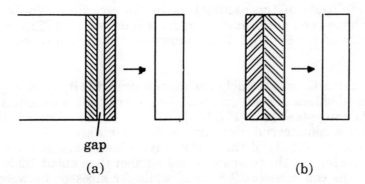

gap

(a)   (b)

Fig. 7: Double impact experiments: a) second shock after complete loading-unloading cycle, b) second shock on top of the first one.

The first configuration consists of a flyer which has two impactor discs separated by a small gap (typically a few tenths of a milimeter) which enables one to follow the second shock after a complete cycle of loading and unloading. This technique was first used by Yaziv *et al* [41] to probe the state of alumina AD85 right after it has been spalled. The extent of the spall layer was determined by observing the time needed for the second shock to traverse the specimen after this complete cycle.

The second double impact configuration was originally used by Asay and Lipkin [42] to observe the state of aluminum specimens by sending two consecutive shocks and observing an elastic wave on the second shock. This technique was used in [43] to load an alumina specimen with two consecutive shocks, below and above the HEL. This resulted in a lower HEL value as compared to that measured by a single shock loading, strengthening the claim of a measureable damage which is induced by the first shock in the specimen.

Recovery experiment
The purpose of recovery plate impact experiments is to find correlations between the observed residual changes in the microstructure (dislocation networks, microcrack density etc.) and the well defined stress pulse which caused these changes. Since any 1D strain experiment is limited by the lateral size, one cannot prevent lateral release waves from entering the specimen and inducing tensile stresses in its central region. With metals, the spall ring technique is used to reduce the magnitude of these stresses and, due to their relatively large tensile strength, large specimens can be recovered intact for microscopic examinations. This technique is not very useful with ceramics which have very small tensile strengths and are bound to shatter upon the introduction of tensile stresses of the order of a few tenths of GPa.

Several techniques have recently been devised in order to reduce the lateral unloading waves, of which the star-shaped flyer of Kumar and Clifton

[44] is probably the most popular, (see Fig. 8).

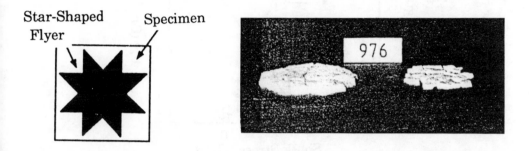

Fig. 8: The star shape impactor (Ref. [44]), and a recovered alumina specimen (Ref. [23]).

The idea here is that the interaction of waves, generated at the edges of the flyer, leave the central region of the impacted specimen almost free from lateral relief waves. These features have been confirmed by Rabie *et al* [45] and Kirpatrick *et al* [46] who performed 3D numerical simulation of the plate impact experiment. This technique was used by several workers who succeeded in recovering spalled specimens from impacts in the elastic range of response [47] and even from high velocity impacts reaching stresses as high as twice the HEL [23]. According to Longy and Cagnoux [25], recovered alumina specimens did not show any damage even when shocked to stresses above their HEL values (see Fig. (8)). This finding is in clear contradiction with others, listed above, which claim that damage to ceramics is induced for stresses well below the HEL. The reason for this discrepancy is not known to us.

Recently, Espinoza *et al* [48] applied 3D simulations and plate impact experiments with star-shaped flyers on alumina and an AlN/Al composite, demonstrating the usefulness of this design for soft recovery experiments. They found that impact stresses up to 2.0 GPa in these materials caused no lateral tensile stresses with microcracks concentrating only at the impact face and on a ring at the back face of the specimen. Both Louro and Meyers [36] and Yeshurun *et al* [37] have demonstrated that the number of microcracks in softly recovered specimens increases considerably when shocked to stresses above about 0.5 HEL. This damage threshold is in good agreement with the onset of decrease in spall strength of these materials (AD85 alumina) as can be seen in Fig. 9.

Shock and Impact on Structures

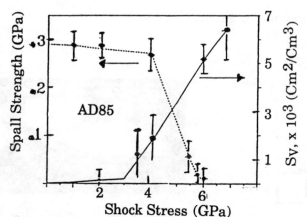

Fig. 9: Reduction in spall strength and increase in microcrack densities as a function of applied stress amplitude (from [37]).

SHEAR STRENGTH AT HIGH DYNAMIC PRESSURES

Considering the fact that during any penetration process the target material ahead of the penetrator is subjected to very high pressures, it is of utmost importance to determine the shear strength of ceramics at these high pressures. Since ceramics are damaged by the first shock which passes through them it is reasonable to assume that the shear strength of the shock loaded material will be higher than that of the relaxed material. Two techniques have been developed to measure these parameters: (i) the lateral stress gauge technique to measure the principal stresses (longitudinal and lateral) in the shocked specimen (from which the shear strength is determined), and (ii) the pressure-shear plate impact test, in which a shear wave is propagated through a specimen which has been loaded by a precursor longitudinal wave.

The lateral stress gauge technique

The only direct technique to measure the shear strength of shock loaded specimens is by embedding two, very thin, piezoresistance gauges in the specimens, one measuring the longitudinal stress σ_x, while the other measures lateral stress σ_y, as shown in Fig. 10 below:

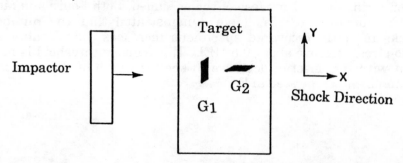

Fig.10 Longitudinal (G1) and (G2) and lateral stress gauges in a plate impact experiment.

The analysis of data obtained with these gauges is relatively simple since the shear stress is equal to half the difference between longitudial and lateral stresses $\tau = \frac{1}{2}(\sigma_X - \sigma_Y)$ as first suggested by Bernstein *et al* [49] who inserted manganin wire transducers in steel specimens. In order to account for the response of the lateral gauge a careful examination of its loading and unloading characteristics was needed. Such a study was the subject of intensive research which is summarised in [50] and the references therein. After calibrating the response of the manganin gauge in the lateral configuration we were able to apply the technique to measure the shear strength of different ceramics at high pressures under plate impact loading. This was repeated for alumina [51] aluminum nitride [52] and titanium diboride [53]. It is interesting to note that while Al_2O_3 and AlN showed a relatively constant $\tau(\sigma_x)$ behaviour (for shock amplitudes above the HEL), the curve for TiB_2 showed a definite increase for stresses well above the HEL. Fig. 11 shows our results for these three ceramics as given in the Refs. [51]-[53].

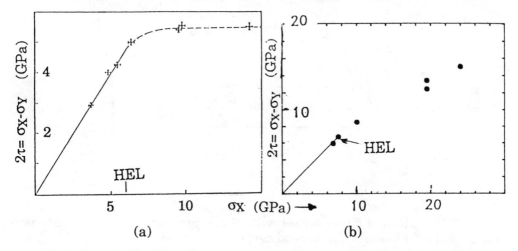

Fig 11: The dependence of shear strength on shock amplitude for 2 ceramics: a) Al_2O_3 [51], and b) TiB_2 [53].

A somewhat indirect technique to determine τ involves considering the offset of the Hugoniot curve from the hydrostatic compression curve at a given volume V. This offset, should be equal to

$$\delta = \sigma_x(V) - P(V) = \frac{4}{3}\tau \qquad (4)$$

so that if both curves are measured accurately one can determine the shear strength at high pressure. The limitation of this technique lies in the fact that P=P(V) is not known accurately enough for most materials, thus, introducing a large error in the values of the shear stress.

This technique was used by Grady [33] who shows (see Fig. (12) below) that SiC and TiB2 are "pressure strengthening", while B4C loses its shear strength when shocked above the HEL. AlN, on the other hand, results in a constant shear strength, in good agreement with our findings [52]. Also, Grady's data on TiB2 agree with ours in [53] which were obtained with lateral stress gauges.

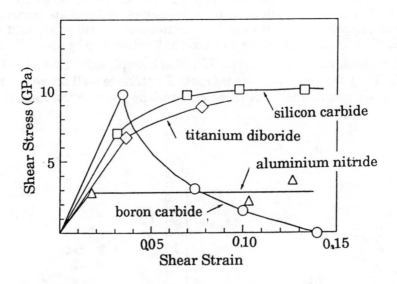

Fig. 12: Variation of shear stress with shock pressure (shear strain), using calculated offset values, for several ceramics (from [33]).

The pressure shear experiment

The high strain rate pressure-shear impact experiment was developed by Clifton and his colleagues [55]-[57] to obtain stress-strain curves for metals at high shear strain rates (above 10^5 s^{-1}). The experiment consists of impacting a thin specimen plate against a hard anvil at an angle θ as shown in Fig. 13. The specimen is backed by another hard disk, which is the flyer plate in the experiment. The flyer and anvil plates remain elastic throughout the experiment and the relatively soft specimen experiences a state of combined shear and high pressure. The stress-strain relation for the specimen is obtained through the measured particle velocity at the rear of the anvil, as shown in the figure.

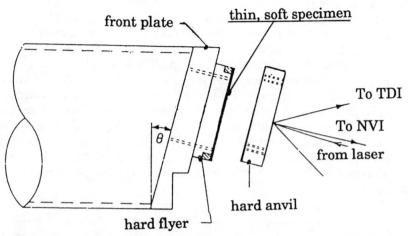

Fig. 13: Configuration of the high pressure-shear experiment.

   The disadvantage of the configuration shown in Fig. 13 is that it requires that the anvil and flyer plates remain elastic throughout the measurement time. This requirement cannot be met if one wishes to investigate the yielding behaviour of specimens like the strong ceramics used for armour. An alternative experiment was used by Espinoza and Clifton [58] and Klopp and Shockey [59], as shown in Fig. 14.

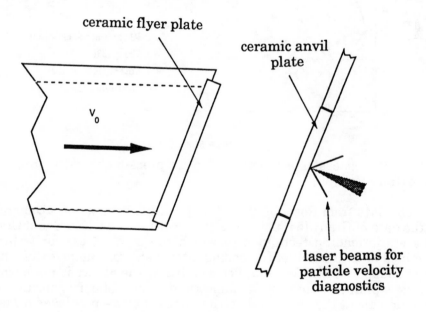

Fig. 14: Symmetric pressure shear experiment.

In this test the impact velocity and angle are varied and the histories of normal and transverse particle velocity are measured at the target rear surface. Here the separation between normal and shear waves is clear and the shear state of the specimen is determined by the transverse velocity history and a material model which is adjusted until the predicted history coincides with the measured one (see [59] for detailed examples on AlN and Al2O3). Figure 15 shows the results for shear stress versus pressure for AlN, as obtained by the various techniques described above. The uppermost data point of Ref. [60] is from a thin specimen impact (Fig. (13)) while the lower point is from a symmetric impact (Fig. (14)). The relatively large difference between the two is the result of the fact that a large hydrostatic pressure is added to the shear in the first configuration, with almost negligible pressure in the second one.

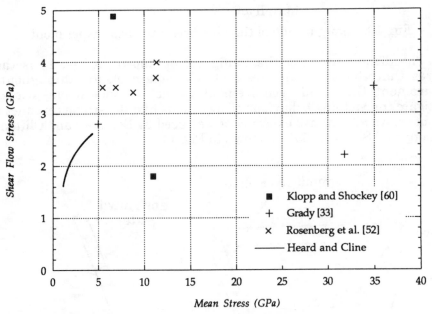

Fig. 15: Shear stress as a function of mean pressure for AlN, from various sources (from Ref. [60]).

The data from Rosenberg *et al* [52] using lateral gauges compare well with the data of Grady [54] using the offset calculations. Heard and Cline [61] used a static loading technique in which the axial stress for failure was determined as a function of increasing lateral stress. The fact that the data point of Klopp and Shockey [60] lies well below the others is not surprising, since, as explained above, the symmetric pressure shear experiment gives information about the shocked material after it has been unloaded. This is so because the shear wave reaches the free surface of the specimen after the normal wave has been reflected from it. Thus, unlike the other technique, this pressure-shear experiment probes the material after a complete loading-unloading cycle. This information is very important for the modelling of ceramics as armour elements because the penetrated tile experiences a complex loading-unloading history before it is reached by the penetrator. Thus it is not enough to characterise the dynamic response of the "virgin" material but, rather, one has to add the constitutive relation for the material which failed after a complete loading-unloading cycle.

MODELLING CERAMIC BEHAVIOUR FOR NUMERICAL SIMULATIONS

Brittle solids can undergo extensive damage when subjected to compressive loading as a result of the high stress concentrations which develop at various irregularities inside their bulk. Analytical modelling of crack nucleation and extension under static loading has been presented by Nemat-Nasser and Horii [62][63] and by Ashby and his colleagues [64][65], to name a few. Dynamic failure of ceramics under impulsive loading is even more complex because of the extremely short times involved and the difficulty to analyse the damage in recovered specimens, as explained above. Compressive fracture, due to microcracking, degrades both the yield strength and the elastic moduli of the specimen. These issues should be accounted for, together with possible pressure, temperature and strain-rate sensitivities of the physical parameters. Thus, modelling the failure of ceramics for numerical simulations consists of an "elasto-plastic" yield envelope and a failure criterion which brings an element from the envelope to the "failed" condition. The failed material should be described by the complex dependence of yield and moduli on all the aforementioned parameters.

Quite understandably, none of the models currently used has all the features necessary to predict the response of ceramics to a general triaxial stress loading. However, various models have been successful in accounting for limited aspects, in specific experiments, such as the stress and particle velocity histories in plate impact experiments. The article by Rajendran [66] reviews many of these models, with a detailed account of their special features. Rajendran divides the various models into four groups as follows:

1) Instantaneous models, in which failure occurs instantaneously when one of the state variables (stress, strain, energy etc.) reaches a critical value. The calculations of Wilkins in [1]-[4] used such a model with a tensile stress threshold of about 0.3 GPa for the ceramics.

2) Plasticity based models in which the yield strength of the ceramic is analogous to that of metals (either von Mises or Tresca yield criteria) with pressure, temperature, and strain-rate sensitivities. Steinberg [17] and Furlong et al [18] used such models to reproduce the particle velocity histories, measured by Grady and Kipp [22], for several ceramics. Figure 15 shows the excellent agreement between these model predictions and the experimental record for SiC.

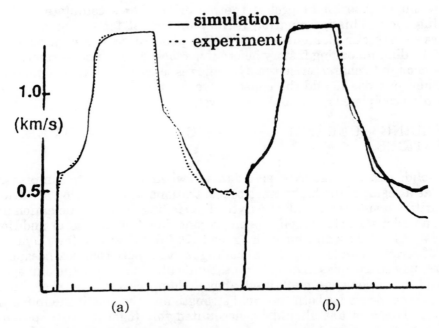

Fig. 16: The agreement between experiment and simulations of a particle velocity record. a) Steinberg [17] and b) Furlong *et al* [18].

Steinberg's yield strength function has the form:

$$Y = \left(Y_A + D.\dot{\varepsilon}^n\right)\frac{G(P,T)}{G(o)} \qquad (5)$$

where G is shear modulus and D, n, Y_A are material parameters. Y_A is actually specimen dependent since it may vary with grain size, porosity, or mechanical history. The rate dependent part $D.\dot{\varepsilon}^n$ is limited by another material constant Y_L for very high strain rates.

As already mentioned, the most serious disadvantage of these models is the fact that they ascribe elasto-plastic properties to brittle materials. This introduces various difficulties in describing the different behaviour of intact and comminuted materials. In order to overcome these difficulties, at least partially, a new generation of elasto-plastic models have recently been presented, as the Johnson Holmquist model [67]. In this model the failure envelope of the ceramic has features which resemble their inherent properties (much higher compressive strength, increase of shear strength with confining pressure etc.). A material element which is brought to this yielding envelope is then automatically moved to the failure curve which lies below the envelope (decreased strength) by a certain factor. This failure curve accounts for the comminuted material ahead of the penetrator and its parameters are taken to be those of the Mohr-Coulomb criterion. Figure 17 shows this model as taken from [67].

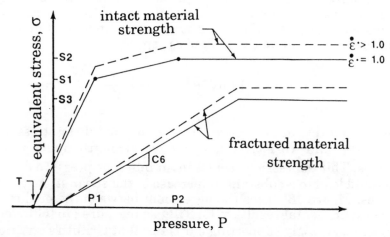

Fig. 17: The Johnson-Holmquist model for a ceramic material (from [67]).

3) Nucleation and growth models have been developed by SRI researchers for both metals and brittle materials [68]-[70]. These models assume that the specimen has an inherent distribution of flaws (inclusions, voids, grain boundaries etc.) which are nucleation sites for failure. Nucleation and growth rates are defined with material parameters which are then calibrated by both plate impact and recovery experiments. The final stage of these models treats the coalescence of cracks and voids leading to predictions of material separation and fragmentation.

4) Fracture mechanics based models, which consider the stress concentrations at crack tips and use failure criteria similar to that of Griffith for brittle solids. The bedded crack model of Margolin [71][72] is one of the most popular models whose basis is a generalised Griffith criterion for crack growth, under both compressive and tensile loading. The sliding mode of crack growth is due to frictional forces across crack faces under compression. The idea of using Griffith's failure criteria instead of that of von Mises (or Tresca), for the simulations of penetration into ceramics, has been demonstrated by Mescall and Tracy [8] who implemented the simple Griffith criterion in the form:

$$(\sigma_1 - \sigma_2)^2 - 8Y_0(\sigma_1 + \sigma_2) = 0 \tag{6}$$

where σ_1, σ_2 are principal stresses and Y_o is the tensile yield stress of the specimen. Their code simulations clearly show the different failure zones, by compression and tension, ahead of the penetrating projectile.

We have already pointed out the fact that by using Griffith's criterion (Eq. (6)), instead of von Mises or Tresca criteria, we were able to obtain a much better agreement between the measured and predicted HEL values for ceramics (based on their static compressive strengths). Carrying this idea a step further we showed recently [73] that the Griffith criterion results in the following expression for the shear stress as a function of pressure behind a planar shock wave above the HEL of the specimen:

$$\tau^* = \frac{1}{12}\left[1 + 6\sqrt{2P^*}\right] \qquad \tau^* = \frac{\tau}{Y_o}; \qquad P^* = \frac{P}{Y_o} \qquad (7)$$

where τ^* and P^* are the normalised values for shear stress and pressure, respectively, and Y_o is the compressive strength under uniaxial stress conditions. This expression results in an inherent pressure dependence for τ which could lead to a substantial increase of the shear strength at high shock pressures. Figure 18 shows the agreement between the prediction of Eq. (7) and the experimental results of for TiB2 as measured in [53]. The agreement is very good, strongly supporting our claim that Griffith's criterion should be used throughout the whole process of dynamic loading of brittle materials.

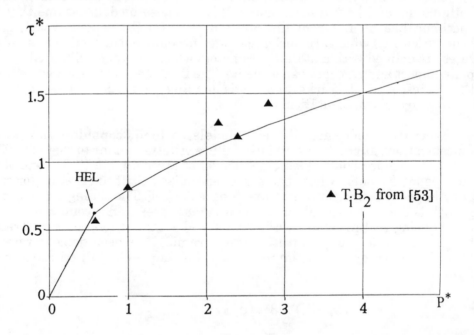

Fig.18: The agreement between predicted and measured shear strength dependence on shock pressure for TiB2.

This treatment of the failure envelope of shock loaded brittle solids can also account for another experimental observation which concerns the shape of the measured stress and particle-velocity histories in these materials. As we have already mentioned above, with rising stress amplitude we obtain (i) a square wave below the HEL, (see Fig. (2)), (ii) an elastic jump to the HEL followed by a ramp, when the stress is up to about 1.5 HEL (see Fig. (3)), and (iii) an S shaped trace for the deformation wave which corresponds to the final shock amplitude for higher stresses (as in Fig. (4)). These features must stem from the shape of the Hugoniot curves of brittle materials.

Now, since there is a simple relation between the offset (S) of the Hugoniot from the hydrostat and the shear strength of the shocked material, we may conclude that the general shape of the $\delta=\delta(P)$ curve will be similar to that of $\tau^*=\tau^*(P^*)$ which we found above (Eq. (7) and Fig. (18)).

In order to construct the Hugoniot curve for any material we have to add the hydrostat and the offset curves, according to the simple relation $\sigma_X=P+\delta$. Since most brittle materials have very high elastic moduli we may assume that their hydrostats are fairly linear at the lower stress range (up to about 2 HEL). Thus the shape of the Hugoniot curve in the low stress range should be similar to that of the $\tau=\tau(P)$ curve while for higher stresses the nonlinearity of the bulk moduli should stiffen it up, changing its curvature, as shown schematically in Fig. 19 below.

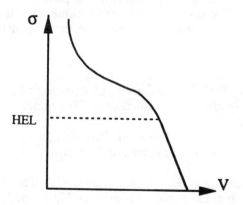

Fig. 19: The schematic shape of the Hugoniot curve for brittle solids

Such a shape for the Hugoniot curve can account for the stress and particle velocity histories which were observed for various ceramics at different shock amplitudes, as seen in Figs. (2) - (4).

CONCLUDING REMARKS

The dynamic response of strong ceramics to shock wave loading has been reviewed by summarising the main findings of various workers over the past two decades. These materials are characterised by very high Hugoniot elastic limits (6 - 20 GPa) and extremely low spall strengths (0.3 - 0.6 GPa). There are strong indications that damage is induced in the shock loaded ceramic even when shocked to relatively low stresses (about half its HEL). Various failure criteria, which are used in numerical simulation codes, have been discussed in an attempt to account for their advantages and discrepancies. It seems that much of the controversy, concerning the high HEL values of ceramics, can be eliminated by choosing Griffith's failure criterion instead of those of von Mises or Tresca, which were developed for ductile yielding. We have shown that the use of this criterion can account for (i) the very high HEL of shock loaded ceramics; (ii) their "pressure hardening" property according to which the shear strength increases with shock pressure, and (iii) the general shapes of the measured stress and particle-velocity histories of these materials, at the different regimes below and above the HEL.

Since the ballistic efficiency of a given armour material is directly related to its shear strength, it is reasonable to assume that those ceramics which exhibit a strong "pressure hardening" effect (such as TiB_2 and S_iC) should perform well, when properly confined. Preliminary data on confined TiB_2 tiles [74] show that their ballistic efficiency, against long-rod penetrators, increases substantially when confined in a strong steel casing. Thus, the knowledge gained by performing shock wave experiments on ceramics, may help the designers in improving their armour sytems by using the right materials and by optimizing geometrical considerations.

REFERENCES
1. Wilkins, M.L., Honodel, C.A. and Sawle, D. Approach to the Study of Light Armour, Lawrence Radiation Laboratory, Livermore, California, Report No. UCRL-50284, 1967.
2. Wilkins, M.L., Cline, C.F. and Honodel, C.A. 4th Progress Report on Light Armour Program, Lawrence Livermore National Laboratory Report No. UCRL-50694, 1969.
3. Willkins, M.L., Landingham, R.L. and Honodel, C.A. 5th Progress Report on Light Armour Program, Lawrence Livermore National Laboratory Report No. UCRL-50980, 1971.
4. Wilkins, M.L. Use of Boron Compounds in Lightweight Armour, in Boron and Refractory Borides, (Ed. Matkovich, V.I.), pp 633-648, Springer-Verlag, 1977.
5. Gust, W.H., Holt, A.C. and Royce, E.B. Dynamic Yield Compressional, and Elastic Parameters for Several Lightweight Intermetallic Compounds, J.Appl.Phys., Vol.44, pp. 550-560, 1973.
6. Gust, W.H. and Royce, E.B. Dynamic Yield Strengths of B_4C, BeO, and

Al2O3 Ceramics, J.Appl.Phys., Vol.42, pp. 276-295, 1971.

7. Gust, W.H. Hugoniot Elastic Limits and Compression Parameters for Brittle Materials, in High Pressure Science and Tech. Proceedings of the VII AIRAPT Conf. Le Creusot, France, p. 1009, 1979.

8. Mescall, J. and Tracy, C. Improved Modeling of Fracture in Ceramic Armor, Proceedings of the 1986 Army Science Conference, U.S. Military Academy, West Point, 17-20 June, 1986.

9. Chartagnac, P. A Pragmatic Approach to Sequential Dynamic Loading, in Shock Compression of Condensed Matter, (Eds. Schmidt, S.C., Johnson, J.N. and Davison, L.W.), p.923, Elsevier Science Publisheres, 1990.

10. Shockey, D.A., Marchand, A.H., Skaggs, S.R., Cort, G.E., Burkett, M.W. and Parker, R. Failure Phenomenology of Confined Targets and Impacting Rods, Int. J. Impact Eng., Vol. 9(3), p. 1, 1989.

11. Forrestal, M.J. and Longcope, D.B. Target Strength of Ceramic Materials for High-Velocity Penetration, J.Appl.Phys., Vol.67, pp. 3669-3672, 1990.

12. Munson, D.E., and Lawrence, R.J. Dynamic Deformation of Polycrystalline Alumina, J. Appl. Phys., Vol. 50, pp. 6272-6282, 1979.

13. Kipp, M.E., and Grady, D.E. Elastic Wave Dispersion in High Strength Ceramics, in Shock Compression in Condensed Matter, (Eds. Schmidt, S.C., Dick, R.D., Forbes, J.W. and Tasker, D.G.), p. 459, Elsevier Science Publishers, 1992.

14. Rosenberg, Z., Yeshurun, Y. and Brandon, D.G. Dynamic Response and Microstructure of Commercial Alumina, J. Phys. France Colloq., Vol.46-C5, pp. 331-342, 1985.

15. Cagnoux, J. and Longy, F. Is the Dynamic Strength of Alumina Rate-Dependent?, in Shock Waves in Condensed Matter 1987, (Eds. Schmidt, S.C. and Holmes, N.C.), pp. 293-296, published North Holland, Amsterdam, 1988.

16. Brar, N.S., Bless, S.J. and Rosenberg, Z. Brittle Failure of Ceramic Rods under Dynamic Compression, J. Phys. France Colloq., Vol.49-C3, pp. 607-612, 1988.

17. Steinberg, D. Computer Studies of Dynamic Strength of Ceramics, Lawrence Livermore National Laboratory Report UCRL-ID-106004, September, 1990.

18. Furlong, J.R., Alme, M.L. and Davis, J.F. Modellng the Dynamic Load/Unload Behaviour of Ceramics Under Impact Loading, RDA Report DAALOL3-89-C-0019, July, 1990.

19. Rosenberg, Z. On the Relation Between the Hugoniot Elastic Limit and the Yield Strength of Brittle Materials, J. Appl. Phys., Vol. 74, p. 752, 1993.

20. Rosenberg, Z., Brar, N.S. and Bless, S.J. Elastic Precursor Decay in Ceramics as Determined with Manganin Strain Gauges, J. Phys. France Colloq., Vol. 49-C3, pp. 707-712, 1988.

21. Brar, N.S., Rosenberg, Z. and Bless, S.J. Applying Steinberg's Model to the Hugoniot Elastic Limit of Porous Boron Carbide Specimens, in Shock Compression of Condensed Matter 1991, (Eds. Schmidt, S.C., Dick, RD., Forbes, J.W. and Tasker, D.G.), p. 467, Elsevier Science Publishers, 1992.

22. Kipp, M.E. and Grady, D.E. Shock Compression and Release in High Strength Ceramics, Sandia Report, SAND89-1461, UC-70h, July 1989.

23.Longy, F. and Cagnoux, J. Plasticity and Microcracking in Shock-Loaded Alumina, J. Amer. Ceram. Soc., Vol. 72, pp. 971-979, 1989.

24.Rosenberg, Z. The Response of Ceramics to Shock Loading, in Shock Compression of Condensed Matter 1991, (Eds. Schmidt, S.C., Dick, R.D., Forbes, J.W. and Tasker, D.G.), p. 439, Elsevier Science Publishers, 1992.

25. Mashimo, T. Shock Compression Studies on Ceramic Materials in Shock Waves in Material Sciences, (Ed. Sawaoka, A.B.), p. 113, Springer-Verlag, Tokyo, 1993.

26. Ahrens, T.J., Gust, W.H. and Royce, E.B. Material Strength Effect, in the Shock Compression on Alumina, J. Appl. Phys., Vol. 39, pp. 4610-4616, 1968.

27. Graham, R.A. and Brooks, W.P. J. Phys. Chem. Solids, Vol. 32, p. 2311, 1971.

28. Ahrens, T.J. and Linde, R.K. Response of Brittle Solids to Shock Compression, in Behaviour of Dense Media Under High Dynamic Pressures, Proceedings of the IUTAM Conf. Paris, 1967, p. 325, Gordon and Breach, 1968.

29. Mashimo, T., Hanaoka, Y. and Nagayama, K. Elastoplastic Properties Under Shock Compression of Al_2O_3 Single Crystals and Polycrystals, J. Appl. Phys., Vol. 63, p. 327, 1988.

30. Grady, D.E. and Kipp, M.E. Dynamic Fracture and Fragmentation in High Pressure Shock Compression of Solids, (Eds. Asay, J.R. and Shahinpoor, M.), p. 265, Springer-Verlag, 1992.

31. Taylor, J.W. Experimental Methods in Shock Wave Physics, in Metallurgical Effects at High Strain Rates, (Eds. Rhode, R.W., Butcher, B.M., Holland, J.R. and Karnes, C.H.), pp. 107-128, Plenum Press, New York, 1973.

32. Meyers, M.A. and Aimone, C.T. Dynamic Fracture (Spalling) of Metals, in Progress in Material Science, Vol. 29, pp. 1-96, Pergamon Press, 1983.

33. Grady, D. Shock Wave Properties of High Strength Ceramics, in Shock Compression of Condensed Matter 1991, (Eds. Schmidt, S.C., Dick, R.D., Forbes, J.W. and Tasker, D.G.), p. 455, Elsevier Science Publishers, 1992.

34. Dandekar, D.P. and Benfanti, D. Strength of Titanian Diboride Under Shock Wave Loading, J. Appl. Phys., Vol. 73, p. 673, 1993.

35. Winkler, W.D. and Stilp, A.J. Spallation Behaviour of TiB_2, SiC, and B_4C Under Planar Impact Tensile Stresses, in Shock Compression of Condensed Matter 1991, (Eds. Schmidt, S.C., Dick, R.D., Forbes, J.W. and Tasker, D.G.), p. 475, Elsevier Science Publishers, 1992.

36. Louro, L.H.L. and Meyers, M.A. Effect of Stress State and Microstructural Parameters on Impact Damage of Alumina Based Ceramics, J. Mater. Sci., Vol. 24, pp. 2516-2532, 1989.

37. Yeshurun, Y., Brandon, D.G. and Rosenberg, Z. Impact Damage Mechanisms in Debased Alumina, in Impact Loading and Dynamic Behaviour of Materials, (Eds. Chiem, C.Y., Kunze, H-D. and Meyer, L.W.), pp. 399-406, DGM Informationgesellschaft mbH, Orberursel, Germany, 1988.

38. Clifton, R.J., Raiser, G., Ortiz M. and Espinoza, H. A Soft Recovery

Experiment for Ceramics, in Shock Compression of Condensed Matter 1989, (Eds. Schmidt, S.C., Johnson, J.N. and Davidson, L.W.), pp. 437-440, Elsevier, Amsterdam, 1990.

39. Murr, L.E. Residual Microstructure - Mechanical Property Relationships in Shock Loaded Metals and Alloys, in Shock Waves and High Strain Rate Phenomena in Metals and Alloys, (Eds. Meyers, M.A. and Murr, L.E.), p. 607, Plenum Press, New York, 1981.

40. Grey, G.T. Influence of Shock-Wave Deformation on the Structure/Property Behaviour of Materials, in High Pressure Shock Compression of Solids, (Eds. Asay, J.R. and Shahinpoor, M.), p. 187, Springer-Verlag, 1992.

41. Yaziv, D., Bless, S.J. and Rosenberg, Z. Study of Spall and Recompaction of Ceramics Using a Double Impact Technique, J. Appl. Phys., Vol. 58, p. 345, 1985.

42. Asay, J.R. and Lipkin, J. A Self Consistent Technique for Estimating the Dynamic Yield Strength of a Shock Loaded Material, J. Appl. Phys., Vol. 49, p. 4242, 1978.

43. Rosenberg, Z., Brar, N.S. and Bless, S.J. Determination of the Strength of Shock Loaded Ceramics Using Double Impact Techniques, in Shock Compression of Condensed Matter 1989, (Eds. Schmidt, S.C., Johnson, J.N. and Davidson, L.W.), pp. 385-388, North-Holland, Amsterdam, 1990.

44. Kumar, P. and Clifton, R.J. A Star Shaped Flyer for Plate-Impact Recovery Experiments, J. Appl. Phys., Vol. 48, pp. 4850-4852, 1977.

45. Rabie, R.L., Vorthman, J.E. and Dienes, J.K. Three Dimensional Computer Modeling of a Shock Recovery Experiment, in Shock Wave in Condensed Matter 1983, (Eds. Asay, J.R., Graham, R.A. and Straub, G.K.), p. 199, Elsevier Science Publishers, 1983.

46. Kirkpatrick, S.W., Curran, D.R., Ehrlich, D.C. and Klopp, R.W. Three Dimensional Analyses of Plate Impact Experiments with Circular and Star Geometries, in Shock Compression of Condensed Matter 1991, (Eds. Schmidt, S.C., Dick, R.D., Forbes, J.W. and Tasker, D.G.), pp. 935-938, Elsevier Science Publishers, 1992.

47. Yaziv, D., Bless, S.J. and Rosenberg, Z. Shock Fracture and Recompaction of Ceramics, in Shock Waves in Condensed Matter, (Ed. Gupta, Y.M.), pp. 425-430, Plenum Press, New York, 1986.

48. Espinoza, H.D., Raiser, G., Clifton, R.J. and Ortiz, M. Performance of the Star-Shaped Flyer in the Study of Brittle Materials: Three Dimensional Computer Simulations and Experimental Observations, J. Appl. Phys., Vol. 72, p. 3451, 1992.

49. Bernstein, D., Godfrey, C., Klein, A., and Shimmin, W. Research on Manganin Pressure Transducers, in Behaviour of Dense Media Under High Dynamic Pressures, Proceedings of the IUTAM Symposium, Paris, 1967, p. 461, Gordon and Breach, 1968.

50. Rosenberg, Z. and Brar, N.S. The Influence of the Elasto-Plastic Properties of Piezoresistance Gauges on Their Loading-Unloading Characteristics as Lateral Stress Tranducers, sent for publication to J.A.P.

51. Rosenberg, Z., Yaziv, D., Yeshurun, Y. and Bless, S.J. Shear Strength of

Shock-Loaded Alumina as Determined with Longitudinal and Transverse Manganin Gauges, J. Appl. Phys., Vol. 62, pp. 1120-1122, 1987.

52. Rosenberg, Z., Brar, N.S. and Bless, S.J. Dynamic High-Pressure Properties of AlN Ceramic as Determined by Flyer Plate Impact, J. Appl. Phys., Vol. 70, pp. 167-171, 1991.

53. Rosenberg, Z., Brar, N.S. and Bless, S.J. Shear Strength of Titanium Diboride Under Shock Loading, in Shock Compression of Condensed Matter 1991, (Eds. Schmidt, S.C., Dick, R.D., Forbes, J.W. and Tasker, D.G.), p. 471, Elsevier Science Publishers, 1992.

54. Chhabildas, L.C. and Swegle, J.W. Dynamic Pressure-Shear Loading of Materials Using Anisotropic Crystals, J. Appl. Phys., Vol. 51, p. 4799, 1980.

55. Rosenberg, Z. and Bless, S.J. Possibility of Measuring Shear Waves in Oblique Impact Experiments with In-Material Piezoresistance Gauges, J. Appl. Phys., Vol. 59, p. 3928, 1986.

56. Gupta, Y.M. Shear and Compression Wave Measurements in Shocked Polycrystalline Al_2O_3, J. Geophys. Res., Vol. 88, p. 4304, 1983.

57. Abou-Sayed, A.S., Clifton, R.J. and Herman, L. The Oblique Plate Impact Experiment, Experimental Mechanics, Vol. 16, p. 127, 1976.

58. Espinoza, H.D. and Clifton, R.J. Plate Impact Experiments for Investigating Inelastic Deformation and Damage of Advanced Materials, A.M.D., Vol. 130, p. 37, 1991.

59. Klopp, R.W. and Shockey, D.A. The Strength Behaviour of Granulated Silicon Carbide at High Strain Rates and Confining Pressures, J. Appl. Phys., Vol. 70, p. 7318, 1991.

60. Klopp, R.W. and Shockey, D.A. Tests for Determining Failure Criteria of Ceramics Under Ballistic Impact, SRI Report ADA 256652, June 1992.

61. Heard, H.C. and Cline, C.F. Mechanical Behaviour of Polycrystalline BeO, Al_2O_3 and AlN at High Pressure, J. Mater. Sci., Vol. 15, pp. 1889-1897, 1980.

62. Horii, H. and Nemat-Nasser, S. Overall Moduli of Solids with Microcracks: Load-Induced Anisotropy, J. Mech. Phys. Solids, Vol. 31, pp. 155-171, 1983.

63. Horii, H. and Nemat-Nasser, S. Brittle Failure in Compression: Splitting, Faulting and Brittle-Ductile Transition, Phil. Trans. R. Soc., Vol. A319, pp. 337-374, 1986.

64. Ashby, M.F. and Hallam, S.D. The Failure of Brittle Solids Containing Small Cracks under Compressive Stress States, Acta. Metal., Vol. 34, pp. 497-510, 1986.

65. Sammis, C.G. and Ashby, M.F. The Failure of Brittle Porous Solids Under Compressive Stress States, Acta Metal. Vol. 34, p. 511, 1986.

66. Rajendran, A.M. and Cook, W.H. A Comprehensive Review of Modeling of Impact Damage in Ceramics, University of Dayton Report UDRI-TR-88-125, October, 1988.

67. Johnson, G.R. and Holmquist, T.J. A Computational Constitutive Model for Brittle Materials Subjected to Large Strains, High Strain-Rates and High Pressures, in Shock Wave and High Strain Rate Phenomena in

Materials, (Eds. Meyers, M.A., Murr, L.E. and Staudhammer, K.P.), p. 1075, Marcel Dekker, 1992.

68. Curran, D.R., Seaman, L. and Shockey, D.A. Linking Dynamic Fracture to Microstructural Processes, in Shock Waves and High Strain Rate Phenomena in Metals, (Eds. Meyers, M.A., and Murr, L.E.), p. 129, Plenum Press, New York, 1981.

69. Curran, D.R., Seaman, L. and Shockey, D.A. Dynamic Failure in Solids, Phys. Rep., Vol. 147, p. 253, 1987.

70. Curran, D.R., Seaman, L. and Cooper, T. A Micromechanical Model for Granular Material and Application to Ceramic Armor, in Shock Compression of Condensed Matter 1991, (Eds. Schmidt, S.C., Dick, R.D., Forbes, J.W. and Tasker, D.G.), p. 551, Elsevier Science Publishers, 1992.

71. Margolin, L.G. A Generalised Griffith Criterion for Crack Propagation, Eng. Fract. Mech., Vol. 19, p. 539, 1984.

72. Margolin, L.G. Numerical Simulation of Fracture, in Proceedings of Int. Conf. on Constitutive Laws for Engineering Materials, (Eds. Desai, C.S. and Gallagher, R.H.), p. 567, Tuscon Arizona, 1983.

73. Rosenberg, Z. On the Shear Strength of Shock Loaded Brittle Solids, sent to J.A.P.

74. Bless, S.J., Benyami, M., Apgar, L.S. and Eylon, D. Impenetrable Ceramic Targets Struck by High-Velocity Tungsten Long-Rods, Presented at the 2nd Conf. on Structures Under Shock and Impact (SUSI), Portsmouth UK, June 16-18, 1992.

APPENDIX - BASICS OF SHOCK WAVE PHYSICS

In this Appendix we briefly summarise the basic features of shock wave physics. The interested reader is referred to the review articles [A1]-[A5] which cover this subject in a comprehensive way.

Shock wave processes are encountered when materials are subjected to impulsive loading with a very short application time, compared to the time needed to respond inertialy. The creation of a shock wave is a direct consequence of the fact that, for most materials, sound speed increases with compression. Thus, an impulsive load applied to the surface of a body will steepen up as it propagates into its interior, as shown schematically in Fig. (A1). This steepening-up is a direct consequence of the fact that the compression curve, for most solids, looks like the one shown schematically in Fig. (A1). Thus, higher pressure increments will move at a higher speed in the solid resulting in the creation of a shock front.

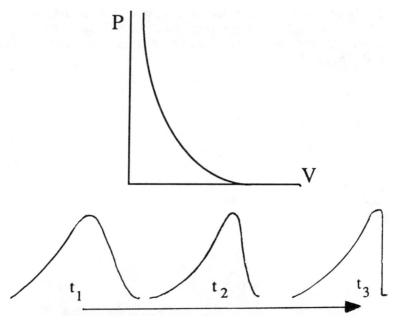

Fig (A1): Schematics of shock wave formation.

The physical process of compression by a shock is adiabatic in nature, increasing both the temperature and entropy of the shocked material. The shock front moves at a velocity (Us) which is determined by its amplitude, imparting a particle velocity (Up) to each material point which is swept by this extremely rapid (almost discontinuous) transition. From both experimental and analytical points of view it is easiest to work with 1-D shock waves which are realised behind planar fronts. The material is then compressed only in the shock propagation direction under uniaxial strain loading.

The mathematical treatment of shock waves, for fluid-like materials, was developed by Rankine and Hugoniot in the previous century and is summarised by the three conservation equations across the front of a plane shock wave - the Rankine-Hugoniot equations:

mass:

$$\rho_o Us = \rho(Us - Up) \qquad (A\text{-}1)$$

momentum:

$$P - Po = \rho_o Us Up \qquad (A\text{-}2)$$

energy:

$$E - E_o = \left(\frac{P + P_o}{2}\right)(Vo - V) \qquad (A\text{-}3)$$

where $V = 1/\rho$ is specific volume and the subscript 0 stands for the undisturbed material ahead of the shock front.

The R-H curve of a given material is built by a series of experiments, in which two of the variables appearing in A1-A3 are measured, so that each shocked state is characterised by the five physical/mechanical parameters: Us,Up, ρ P, E. It was found empirically that for most materials a linear relationship exists between Us and Up according to:

$$Us = Co + SUp \qquad (A\text{-}4)$$

where Co is the bulk sound speed (at zero pressure) of the material and S ranges between 1-1.5 for most materials. When presenting shock data it is convenient to work with either Us=Us(Up) or with P=P(Up) representations which, according to Eq. (A-2) and (A-4), take the forms shown in Fig. (A2).

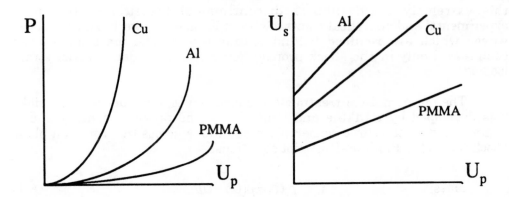

Fig. (A2): Schematics of shock Hugoniot curves for three different materials.

Plane shock wave experiments are conducted with either explosive lenses in contact with a disc shaped specimen, or by plate impact techniques using laboratory guns to accelarate a flyer plate to the desired velocity towards a stationary specimen disc, as shown in Fig. A3 below. The amplitude of the shock waves, which sweep the flyer and target plates, is determined by the impact velocity and their respective Hugoniot curves.

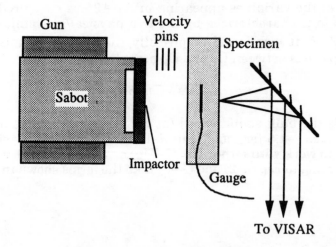

Fig. (A3): A plate impact experiment.

The Hugoniot curves for solids are different fron those of fluids because of the finite strength which solids possess. If the impact velocity is low enough an elastic shock wave is transmitted through the specimen while above a certain threshold velocity, dynamic yielding will take place. Yielding under dynamic compression is manifested by a break in the Hugoniot curve which for low pressures starts with an elastic (straight) line according to the elastic equations:

$$\sigma_X = \left(K + \frac{4}{3}G\right)\varepsilon_X \qquad \varepsilon_X = \varepsilon_V$$

$$\sigma_Y = \sigma_Z = \frac{v}{1-v}\sigma_X \qquad\qquad (\text{A-5})$$

where K and G are the bulk and shear moduli, respectively, and $\sigma_X, \sigma_Y, \sigma_Z$ are the stresses in the shocked specimen, along the principal coordinate system. Note that $\varepsilon_X = \varepsilon_V$ is the only non-vanishing strain component. σ_X is the longitudinal stress behind the shock front while σ_Y, σ_Z are the lateral stresses.

These elastic relations hold until the maximum shear stress in the specimen reaches its limiting value which, according to either von Mises or Tresca criteria, is equal to half the yield strength (Y) of the specimen. Thus, one can write:

$$\tau = \frac{1}{2}(\sigma_X - \sigma_Y) = \frac{1}{2}\left(\frac{1-2v}{1-v}\right)\sigma_X = \frac{1}{2}Y \qquad\qquad (\text{A-6})$$

from which the maximum value of σ_X on the elastic part of the Hugoniot curve (the Hugoniot Elastic Limit) is obtained as:

$$HEL = \frac{1-v}{1-2v}Y \qquad\qquad (\text{A-7})$$

A Hugoniot curve for a real solid will have a cusp at the HEL leading to a splitting of shock waves with amplitude larger than the HEL. Fig. (A-4) shows such a splitting, together with a Hugoniot curve for a so-called elasto-plastic solid.

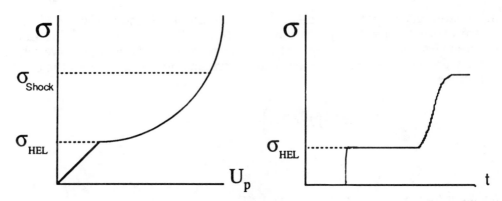

Fig. (A-4): The Hugoniot curve for a material with strength and the double wave caused by the cusp in the Hugoniot.

Numerous techniques have been developed over the past 40 years in order to probe the state of the shocked material. The first efforts concentrated on the Us-Up lines in order to determine a data base for the Hugoniot curves of various solids and liquids. New techniques were developed in the late 60's with which time resolved stress and particle velocities were measured at selected locations inside the shocked specimen. Of these, the most popular are the VISAR interferometry for accurate particle velocity measurements and the embedded manganin stress gauge, with which stresses up to 1.5 Mbar (150 GPa) have been measured by SRI workers. These time resolved measurements enabled experimentalists to explore various dynamic phenomena like yielding, phase transitions etc. These measurements were linked with the dynamics of dislocation motion in metals through phenomena like elastic precursor decay, in which the elastic wave's amplitude decreases with specimen thickness, and the nonsteady nature of the plastic wave, which is manifested by the increasing rise time of this wave. Some of these features were also observed   with the aid of time resolved measurements on ceramics, as described in the text.

REFERENCES

A1)   Rice, M.H., McQueen, R.G. and Walsh, J.M. Compression of Solids by Strong Shock Waves, in Solid State Physics, Vol. 6, (Eds. Sietz, F. and Turnbull, D), pp. 1-63, Academic Press, New York, 1958.

A2)   Duvall, G.E. Some Properties and Applications of Shock Waves, in Response of Metals to High Velocity Deformation, (Eds. Shewmon, P.G. and Zackey, V.F.), pp. 165-202, Interscience, New York, 1961.

A3)   Duvall, G.E. and Fowles, G.R. Shock Waves, in High Pressure Physics and Chemistry, Vol. 2 (Ed. Bradley, R.S.), pp. 209-291, Academic Press, New York, 1963.

A4)   Altshuler, L.V. Use of Shock Waves in High Pressure Physics, Sov. Phy. Uspekhi, Vol. 8(1), pp. 52-91, 1965.

A5)   Davison, L. and Graham, R.A. Shock Compression of Solids, Phys. Rep., Vol. 55(4), pp. 255-379, 1979.

Chapter 5

Concrete structures under soft impact loads

A. Miyamoto,[a] M.W. King[b]

[a] Department of Civil Engineering, Kobe University, Nada, Kobe 657, Japan

[b] Department of Quality Control Engineering, Sun-Mix Concrete Sdn. Bhd., 11500 Penang, Malaysia

ABSTRACT

The ultimate behaviors and failure modes of reinforced concrete slabs are studied using the layered finite element method and verified through comparisons with results of tests on full scale specimens. Consequently, the analysis is found to be capable of predicting the ultimate behaviors very well. Parametric studies on different conditions are performed. As a result, the impact failure modes for reinforced concrete structures under soft impact loading are found to be related to the loading rate of the impact force-time function. Three distinct failure modes are analytically identified and classified as; bending failure, intermediate failure (bending to punching shear failure), and punching shear failure. Finally, some evaluations of the impact resistance of concrete slabs are performed based on a few indexes. Evaluations of the impact performance are useful for determining the relative impact performances between different types of structures.

1. INTRODUCTION

In recent years, there has been an increased practical application of concrete in various fields of construction activity. The application of concrete has grown largely into areas which were originally dominated by other construction materials, such as steel, mainly because of the economic costs in construction and also maintenance throughout the service life of the structure. Concrete structures have already been in service for some time in structures such as concrete barges and vessels, concrete geothermal power reactor plants, concrete power and diversion tunnels for hydro-electric schemes, prestressed concrete nuclear reactor vessels, concrete containment vessels, concrete nuclear shelters, concrete silos and bunkers, concrete offshore structures, and also liquefied natural gas (LNG) tanks[1.1]. Furthermore, there is also a great possibility of further practical applications, such as for concrete lunar structures[1.2]. But as the scope of application increases, the surrounding environment in which these structures are subjected to becomes more severe and also higher safety requirements in design are required.

Impact or impulsive loading resulting from accidental collisions is a particular case in which a high safety requirement is necessary for structures of public importance or significance. One such example is the accidental collision of an aircraft into a nuclear power reactor[1.3], where failure of the reactor structure would cause leakage of radioactive materials into the atmosphere. The problem of impacts on structures have been gaining attention recently, mainly because of lack of rational and globally accepted design methods or procedures for structures that are susceptible to impact collisions. These can be seen in the recent number of conferences to deal with impact problems[1.4 - 1.15]. Moreover, there is no accepted method of analysis and also evaluation techniques for impact problems. This problem is even more severe when the structures concerned are limited to concrete structures. The ultimate behaviors of concrete structures can be very complex due to the effects of material nonlinearity in concrete, cracking, interaction between concrete and reinforcement(bond), shear transfer at cracked sections, etc. Moreover, the dynamic aspects of the various problems arise during impact loading.

At present, concepts of "*equivalent dynamic loads*" are usually employed during design of concrete structures subjected to impact loads as designers are more accustomed to static design methods. A substitutive method of employing equivalent dynamic loads to static design methods would not be adequate in dealing with the real impact phenomenon. For example, a concrete structure may be designed to withstand the expected maximum impact loads or stresses using static design methods. But the structure might not be able to withstand the effects of excitation of high modes of vibration, a change in failure mode due to dynamic response of structures, scabbing at the rear face of the impacted structure, etc., which are behaviors peculiar to concrete structures under impact loadings[1.16, 1.17, 1.18]. Therefore, it is the author's opinion that a dynamic approach for design of concrete structures would be the only practical solution to this problem. There is also a necessity of evaluating the impact resistance properties of structures based on the structural as well as the material point of view. These are considered as the main aims in the study of impacts on concrete structures.

Prediction of impact failure modes and also the dynamic behaviors at ultimate states are necessary for designing concrete structures that might be subjected to accidental collisions. Furthermore, prevention of total structural failure during an accidental collision is important especially for structures of public significance. Failure of a structural element can also cause severe damage to the entire structure due to the immediate loss of load bearing capacity in the particular element. The main objective of this chapter is to propose a dynamic analytical procedure capable of predicting the ultimate behaviors as well as the failure modes in concrete slab structures subjected to soft impact loads. Consequently, a rational method of design for concrete slab structures subjected to soft impact loads is proposed.

The outline of this chapter can be summarized as follows:

In Chapter 2, numerical modeling of reinforced concrete structures under impact loads is discussed. A multiaxial failure model is chosen to be used together with the plasticity model for concrete while a uniaxial model is selected for reinforcement. The dynamic equation of motion is solved using the *Newmark-β* method together with an iterative procedure. Chapter 3 deals with the impact response of concrete slab structures. The layered finite element method is applied for modeling of reinforced concrete slabs and reinforced concrete guardrails(handrails). Experimental results on full scale concrete slab structures are employed for verifying the analytical procedure. An attempt to identify the different types of failure modes under soft impact loading, i.e., bending failure and punching shear failure, using the analytical procedure is discussed. The dynamic behaviors of concrete slab structures under different conditions are parametrically studied. Finally, the evaluation of impact resistance for both reinforced concrete beams and slabs are proposed based on a few indexes for impact resistance in Chapter 4.

2. NUMERICAL MODELING OF REINFORCED CONCRETE STRUCTURES UNDER SOFT IMPACT LOADS

This section deals with the numerical modeling and representation for analysis of reinforced concrete structures that are subjected to soft impact loads. Firstly, the modeling of the two main constitutive components in a reinforced concrete structure, namely concrete and reinforcement, are considered. Various types of models have been proposed through the years and each model has its own merits and demerits. The plasticity model, which has been around for a long time and also undergone numerous types of alterations in order to express the behavior of materials more appropriately, is mainly considered and applied in the analysis. Different types of models are also introduced, a uniaxial model for modeling of two dimensional plane stress problems and also a triaxial model for modeling of slab structures. A nonuniform hardening model is employed together with a nonassociated flow rule in the plasticity model. A correlation between available uniaxial material characteristics and the effective stress-effective strain relation in the triaxial model is also discussed. A multi-linear uniaxial stress-strain relation is employed for modeling of reinforcements, which could be normal structural steel reinforcements or FRP based reinforcements.

Finally, the dynamic behaviors of reinforced concrete structures are considered and the method employed in this paper is pointed out in detail. The dynamic equation of motion for solving impact problems is discussed. The *Newmark-β* method is employed in this study together with an iterative procedure to enhance the accuracy of the analysis.

2.1 CONSTITUTIVE MODELS FOR CONCRETE AND REINFORCEMENT

It has been well recognized that the incompleteness of material models for reinforced concrete is a major factor limiting the further expansion of analysis on reinforced concrete structures. Most efforts by researchers in this field have been spent on improving the constitutive relations of concrete materials as well as the modeling of interactions between reinforcement and concrete. But due to the complexity of the problems at hand and also the different methods of approach for solving each problem, no uniformed approach for dealing with analysis of reinforced concrete structures has been accepted. On the other hand, the complexity of most available nonlinear analytical procedures has also limited the application of these methods to actual design and practice.

2.1.1 CONSTITUTIVE MODELS FOR CONCRETE MATERIALS

A few modeling approaches have been proposed, namely *nonlinear elasticity, plasticity, endochronic theory, fracture mechanics* and *damage theory*. The plasticity theory will be mainly considered in this study together with the main purpose of application to a dynamic nonlinear finite element program for analysis of reinforced concrete structures subjected to soft impact or impulsive loads.

Compared to the classical theory of plasticity, the endochronic theory [2.1, 2.2] is a new approach towards modeling the constitutive relations of materials. It has the advantage of being able to express the *pre-failure* (*strain hardening*) and *post-failure* (*strain softening*) behavior based on a single equation. This causes no discontinuities in the stress-strain curve, which is very convenient in nonlinear computations. A term, called the *intrinsic time*, is introduced into the model and there is no necessity of dividing the strain into elastic and plastic components as in the plasticity model. The effects of loading and unloading is also relatively easy to handle. The endochronic theory has also been introduced for dealing with the strain rate effects in concrete and reinforcement materials [2.3]. In this case, the model is capable of incorporating not only the yield stress increase with the rate of straining but also an increase of the irreversible deformation up to failure. But these models have the disadvantage of requiring a number of material constants to be specified, thus implying the necessity of relatively accurate test results of stress-strain behavior of concrete materials, which is still a field where much more research and development of test and instrumentation methods are necessary. Furthermore, it is considered here that the effects of the post-failure region which comes into effect only after failure is in a relatively progressed stage, i.e., the strain-softening behavior, is of not much significance in the analysis of transient behaviors, such as soft impact loadings.

1) PLASTICITY MODEL FOR CONCRETE MATERIALS[2.4, 2.5]

Most of the constitutive models proposed to date are of the phenomenal type. In other words, the aim of the phenomenal model is to reproduce mathematically the macroscopic stress-strain relations for different loading conditions, and neglecting the microscopic mechanism of the behavior. The plasticity approach falls into this category. This approach is more convenient when dealing with the global response of total structures or structural components.

A plasticity model has to involve the following three basic assumptions [2.4]:
(1) An *initial yield surface* in stress space defines the stress level at which plastic deformation begins.
(2) A *hardening rule* defines the change of the loading surface as well as the change of the hardening properties of the material during the course of plastic flow.
(3) A *flow rule*, which is related to a *plastic potential function*, gives an incremental plastic stress-strain relation.

In plasticity modeling of concrete, the ultimate strength condition (*failure criterion*) has to be assumed, in addition to the above three assumptions. The theory of plasticity was initially developed for metals and as such considerable modifications have to be carried out to adapt it for application to concrete materials. For metals, the yield criteria is usually applied as the failure criteria, as brittle failure can be usually assumed. In the modeling of concrete, the ultimate strength condition (i.e., the failure criterion) which sets the upper bound of the attainable states of stress has to be assumed in addition to the above three assumptions, as the region between yielding and failure is relatively large.

A proper constitutive model would consist of two main parts, namely the *pre-failure* (*hardening*) range and the *post-failure* (*softening*) range as illustrated in Fig.2.1. As pointed out earlier, the main concern of this study is the application of a proper constitutive model for the analysis of reinforced concrete structures under soft impact or impulsive loading. It is considered that the post-failure region has little significance in the analysis of loading under very short durations, and as such, the scope of this chapter will be limited to the *pre-failure* (*hardening*) region only, i.e., the region up to point "D" in Fig.2.1.

2) FRACTURE MECHANIC MODEL FOR CONCRETE MATERIALS

Fracture mechanics is fast gaining universal attention in the modeling and analysis of reinforced concrete structures. Various reports by different committees have been recently published to deal with the rapid developments that are being seen in this relatively new and growing field [2.6 - 2.9]. Since the failure criterion is energy based in this model, the effects of mesh sensitivity and brittle cracking can be considered and effectively employed to deal with the failure mechanisms in concrete based structures. It is also reported that this model would be suitable for analysis of brittle failure of concrete structures, such as diagonal shear, punching shear, torsion, pull-out and also mass concrete [2.6]. The five main reasons for the rapid growth of fracture mechanics model can be given as;
(1) Energy required for crack formation : Crack initiation may depend on stress or strain but the actual formation of cracks requires a certain amount of energy, which is referred to as the fracture energy (surface energy of the solid).
(2) Objectivity of calculations : This is mainly based on the effects of mesh sensitivity during discrete analysis of concrete structures. This phenomenon can be schematically illustrated in Fig.2.2 for the analysis of crack in a concrete body. Due to the difference in choice of coordinates, meshes, etc., as shown in Fig.2.2(b) and (c), different predictions of the ultimate behaviors (Fig.2.2(d) and (e)) are derived. Furthermore, as illustrated in Fig.2.2(f), application of fracture mechanics allows a same amount of energy to be dispersed without any relation to the amount of elements idealized in a structure.
(3) Lack of yield plateau : The softening part in the stress-strain relations and also progressive failure of concrete through the cross-sections cannot be modeled accurately using the theory of plasticity, especially if the failure is brittle.
(4) Energy absorption capability and ductility : The area enclosed by the load-deflection curve represents the amount of energy absorbed by a structure. This is especially important during dynamic loadings. Since this procedure allows the prediction of the post-peak behaviors, it would provide a better information about the energy dissipated throughout the entire loading process.

(5) Size effect : This phenomenon is illustrated in Fig.2.3. In the figure on the left, test or analysis on specimens of different sizes would produce different stress-relative deflection functions when normal procedures are employed because no size effects are incorporated. Therefore, when the results are extrapolated for actual sized structures, the results would be inaccurate. This is shown in the figure on the right, where lab tests and real structures do not produce matching results. But for fracture mechanics, the size effects can be taken into account.

One of the main reasons for hindering the application of fracture mechanics to concrete structures under impact loading is because of a lack of experimental data for modeling dynamic behavior of concrete by fracture mechanics. Furthermore, the three main brittle failure (cracking) modes that are considered are either Mode I(opening) , Mode II(plane shear) or Mode III(anti-plane shear), as illustrated in Fig.2.4. But in the analysis of concrete structures under impact loading, punching shear in the structure together with the effects of excitation of the higher modes of vibrations and transverse shear stresses and also stress waves during high loading rates, cause the cracks to be of a complicated mixed mode. At present, most research that is being carried out are focused mainly on Mode I type of fracture, which would be adequate for static problems. Some research into the dynamic aspects of the problem have also been carried out recently on Mode I [2.10 - 2.15], Mode III [2.16] and also bond between concrete and reinforcement [2.10]. Since there is no proper model at present to represent such a mixed mode fracture phenomenon, the application to impact problems is still relatively limited. But fracture mechanics should prove to be an effective method of modeling concrete structures under impact loading in the near future.

2.1.2 CONSTITUTIVE MODEL FOR REINFORCEMENT

The main method of modeling of reinforcement in concrete structures is by applying a uniaxial elasto-plastic model. Since reinforcement in concrete are mostly one-dimensional and can be easily modeled by line elements, there is no necessity of introducing complex multiaxial relations for steel and simple models such as multi-linear relations can be employed. Furthermore, the linear stress-strain relations in the elastic stages allow easy representations of the constitutive model. The normal idealizations for structural steel in either compression or tension are usually one of the following models;
(1) Elastic perfectly plastic approximation,
(2) Trilinear approximation,
(3) Complete curve.
In most cases, the stress-strain relationship in tension and also compression are assumed to be identical.

2.1.3 MATERIAL CHARACTERISTICS FOR SOFT IMPACT PROBLEMS [2.17 - 2.19]

Soft impact problems in the field of civil engineering usually involve the collision of highly deformable impacting bodies into large and massive structures. Therefore, the region of strain rate is usually between 10^{-1} to 50 sec^{-1}, with the impact load reaching a maximum value at a range of between 10^{-1} to 10^{-3} seconds. The dynamic aspects of different typical phenomenon are listed in Fig.2.5 [2.20]. The soft impact region would then be the "intermediate strain-rate" shown in the figure. As such, it is considered that the inertia forces would be the main factor contributing to the strain-rate effects. But for higher strain rates, as in hard impacts and hyper-velocity collisions, the adiabatic process would then be of influence and the effects of thermodynamics becomes greater. This is considered to be due to stress waves propagating through the medium and part of the energy would be converted into heat. The different mechanisms by which energy transformation occurs can be collectively termed as internal friction [2.20].

The necessary dynamic considerations for different loading conditions are given in Table 2.1. Since inertia forces would be the main factor that distinguish soft impact loadings from static loadings, the dynamic effects can be simulated by simply applying static material characteristics into the dynamic equation of motion (equilibrium). Since inertia forces can be simulated in the equation of motion, static material characteristics would be adequate. On the other hand, if dynamic based or strain rate based material characteristics are to be included into the dynamic equation of motion, it would result in an overestimation (double estimation) of the real dynamic contribution of inertia forces.

To demonstrate this point, a finite element analysis of a concrete column (300.0 x 150.0 x 50.0*mm*) under compression of different loading rates is performed. Fig.2.6 shows the average stress-average strain relation for three different cases of loading rate and also the corresponding static stress-strain curve of concrete under uniaxial compression [2.18, 2.21]. The damping constant, h, is set at 10%. The figure shows that the ultimate average stress increases with an increase in loading rate, while the ultimate average strain decreases correspondingly. This is the normal phenomenon noticed during dynamic loading. From the results, it is clear that applying only a static based material characteristics into a dynamic equation of motion would produce results of the dynamic effects, namely inertia forces. Therefore, only static material characteristics will be used throughout this study in the analysis of concrete structures under soft impact loading.

2.2 MULTIAXIAL FAILURE AND YIELD CRITERION FOR CONCRETE

The multiaxial failure and yield criterion is considered here for modeling of concrete in slab structures that are subjected to multiaxial stresses. Normally, a biaxial failure and yield model would suffice for slab structures, but it is considered that the transverse shear stresses during impact loadings are enhanced largely and a multiaxial model would be necessary in order to take into account these stress effects [2.22 - 2.24].

2.2.1 FAILURE AND YIELD CRITERION

A variety of failure criteria have been applied to date to predict the ultimate behavior of concrete materials. Fig.2.7 shows a diagrammatic illustration of the better known failure models [2.5]. From results of concrete strength tests reported in the literature, it is known that concrete is a hydrostatic-pressure-dependent material with curved meridians on the failure surface as illustrated in Fig.2.8(a) [2.5]. This indicates that the hydrostatic pressure produces effects of increasing the shearing capacity in concrete. It is also known that a pure hydrostatic loading will not cause failure to occur. The assumption of isotropy is reasonable in concrete materials and thus produces the 120° period and 60° symmetry in the deviatoric planes shown in Fig.2.8(b).

The more refined models proposed by *Hsieh* et al. [2.4, 2.5, 2.25] (Fig.2.7(f)), *Willam* and *Warnke* [2.4, 2.5] (Fig.2.7(g)) and *Ottosen* [2.4, 2.5, 2.26] (Fig.2.7(h)) are capable of reproducing the main features of the triaxial surface and are also reported to give close estimates of relevant experimental data.

The *Ottosen* model [2.26] will be considered here because it is capable of reproducing the main features of a triaxial failure surface with the least amount of parameters, in comparison with the models proposed by *Hsieh* et al. [2.4, 2.5, 2.25] (Fig.2.7(f)), and *Willam* and *Warnke* [2.4, 2.5] (Fig.2.7(g)). To meet the geometric requirements of the failure surface for concrete materials, *Ottosen* proposed the following criterion involving all three stress invariants I_1, J_2 and θ,

$$f(I_1, J_2, \theta) = \frac{a\, J_2}{(f'_c)^2} + \frac{\lambda \cdot \sqrt{J_2}}{f'_c} + \frac{b \cdot I_1}{f'_c} - 1 = 0 \tag{2.1}$$

where λ is a function of $\cos(3\theta)$,

$$\lambda = \begin{cases} k_1 \cdot \cos\left[\dfrac{1}{3}\cos^{-1}(k_2 \cdot \cos(3\theta))\right] & ;\text{for } \cos(3\theta) \geq 0 \\[3mm] k_1 \cdot \cos\left[\dfrac{\pi}{3} - \dfrac{1}{3}\cos^{-1}(-k_2 \cdot \cos(3\theta))\right] & ;\text{for } \cos(3\theta) < 0 \end{cases} \tag{2.2}$$

f_c': uniaxial compressive strength of concrete,
f_t: uniaxial tensile strength of concrete.

In Eqs.(2.1) and (2.2), a, b, k_1, k_2 are constants. The typical values can be found in Table 2.2. These values have been calibrated from the better known biaxial (by *Kupfer* et al. [2.4, 2.5]) and triaxial (by *Balmer* and *Richart* et al. [2.4, 2.5]) tests results. The values for $\overline{f_t}$ $(= f_t / f_c') = 0.12$ are selected for the analysis in this paper.

The failure surface for concrete is generally expressed as,

$$f(\rho, \sigma_m, \theta) = \rho - \rho_f(\sigma_m, \theta) = 0, \quad |\theta| \leq 60° \tag{2.3}$$

where $\rho = \sqrt{2J_2}$ is the stress component perpendicular to the hydrostatic axis, and $\rho_f(\sigma_m, \theta)$ defines the failure envelope on deviatoric planes and is given by different expressions in different failure models.

The expression for $\rho_f(\sigma_m, \theta)$ in the model proposed by *Ottosen* can be expressed as follows,

$$\rho_f(\sigma_m, \theta) = \frac{1}{2a}\left[-\sqrt{2} \cdot \lambda + \sqrt{2 \cdot \lambda^2 - 8a(3b \cdot \sigma_m - 1)}\right] \tag{2.4}$$

where, $\sigma_m = I_1/3$.

The yield criterion for concrete is usually assumed on the basis of a reduced function of the known failure criterion since it is not easy to measure experimentally the yield stress in concrete. In the earlier plasticity models, the yield surface was assumed to be a proportionally reduced shape of the failure surface as shown in Fig.2.9(a). Later it was found that this assumption is inadequate for concrete materials as the yield surface would then be an open surface and yielding under a pure hydrostatic loading would never occur. The experimental results reported by *Launay* and *Gachon* [2.5] shows that yielding in the hydrostatic plane occurs and an elastic limit was also reported. Furthermore, there have also been reports that a failure cap develops in the direction of hydrostatic pressure when concrete is subjected to blast or shock loading [2.27]. But since this study is mainly in the region of soft impact loadings, especially in concrete slab structures where pure hydrostatic pressures are not induced, the closed cap model will not be considered.

A nonuniform hardening plasticity model proposed by *Han* and *Chen* [2.5, 2.28] is applied here. The yield surface for this model is shown in Figs.2.9(b) and 2.10. The failure surface encloses all the loading surfaces and it serves as a bounding (limiting) surface. The initial yielding surface is assumed to have a closed surface. During hardening, the loading surface expands and changes its shape gradually from the initial yield surface to the failure surface. Each loading surface is characterized by a hardening parameter k_0. An example of the nonuniform hardening plasticity model based on results from a uniaxial compressive test result is shown in Fig.2.11.

The shape of the meridians of the yield surface may be assumed to be as shown in Fig.2.12, which consists of four different parts:

(1) In the tension zone, i.e., $\sigma_m \geq \xi_t$, the yield surface coincides with the failure surface. Assume no plastic deformation up to failure, representing brittle failure.

(2) In the compression-tension mixed zone, i.e., $\xi_t > \sigma_m \geq \xi_c$, a plastic-hardening zone gradually evolves.

(3) In the compression zone with a low confining pressure, i.e., $\xi_c > \sigma_m \geq \xi_k$, the meridian represents a proportionally reduced size of the failure surface.

(4) In the compression zone with a relatively high confining pressure, i.e., $\sigma_m < \xi_k$, the yield surface gradually closes up at the hydrostatic axis and a wide plastic-hardening region is generated.

Based on Fig.2.12, the four different zones can be mathematically divided by the following triaxial zoning criteria,

Tension-tension:

$$\sqrt{J_2} - \frac{1}{\sqrt{3}} \cdot I_1 > 0 \tag{2.5}$$

Tension-compression:

$$\sqrt{J_2} - \frac{1}{\sqrt{3}} \cdot I_1 \leq 0 \text{ and } I_1 \geq 0 \tag{2.6}$$

Compression-tension:

$$\sqrt{J_2} + \frac{1}{\sqrt{3}} \cdot I_1 \geq 0 \text{ and } I_1 < 0 \tag{2.7}$$

Compression-compression:

$$\sqrt{J_2} + \frac{1}{\sqrt{3}} I_1 < 0$$

(2.8)

The surface represented by the two following equations denote the uniaxial tension and compression states, respectively,

$$\sqrt{J_2} - \frac{1}{\sqrt{3}} I_1 = 0 \text{ ; uniaxial tension state}$$

(2.9)

$$\sqrt{J_2} + \frac{1}{\sqrt{3}} I_1 = 0 \text{ ; uniaxial compression state}$$

(2.10)

These points can be used as the reference point when referring(calibrating) the results to uniaxial material tests.

The zoning criteria mentioned above with the parameters ξ_t, ξ_c and ξ_k, can be used as an indication of the various stress states but it is not necessary to obtain exact values for all the parameters. For simplicity, the following conditions are assumed: $\xi_t = f_t$, $\xi_c = \xi_k = -f_c'/3$.

The initial yield function can be formulated by introducing a shape factor "k" into the failure function defined by Eq.(2.3), yielding an expression of the form,

$$f = \rho - k \cdot \rho_f(\sigma_m, \theta) = 0$$

(2.11)

The shape factor "k" is a function of the hydrostatic stress, σ_m, which modifies the failure surface so as to give a proper shape for the initial yield surface.

The relation between the shape factor "k" and the hydrostatic stress σ_m can be calibrated against results from uniaxial material compressive tests by substituting Eq.(2.10) into Eq.(2.11). (Refer Fig.2.10). Eq.(2.10) can also be expressed as,

$$\rho = \sqrt{6} \cdot (-\sigma_m)$$

(2.12)

This brings about the following relation,

$$k = \frac{-2a\sqrt{6} \cdot \sigma_m}{-\sqrt{2} \cdot \lambda + \sqrt{2 \cdot \lambda^2 - 8a(3b \cdot \sigma_m - 1)}}$$

(2.13)

Considering the elastic limit to be $0.3f_c'$, the following elastic limiting value for "k" can be obtained,

$$k = k_y \cong 0.6$$

(2.14)

where "k_y" is the value for the initial yield surface.

2.2.2 PLASTICITY FOR CONCRETE MATERIALS

Since the initial yield surface in the present model has a different shape from that of the failure surface, during hardening, both the size and the shape of the subsequent yield surfaces must vary continuously from the initial yield shape to the final failure shape. Such a hardening rule is nonuniform with respect to the hydrostatic stress.

According to the nonuniform hardening rule [2.5, 2.28], the subsequent yield surface can be expressed in a reduced form similar to the initial yield criterion as,

$$f = \rho - k(k_0, \sigma_m) \cdot \rho_f(\sigma_m, \theta) = 0$$

(2.15)

The shape factor "k" is defined as a function of σ_m as well as the hardening parameter k_0. The parameter k_0 indicates the hardening level, which can take a value between k_y and 1, i.e.,

$$k_y \leq k_0 \leq 1 \tag{2.16}$$

where $k_0 = 1$ indicates that the ultimate stress state has been reached and that the loading surface has finally met the failure surface.

The shape factor "k" for all stress states can be defined as in the following function (Refer Fig.2.12),

$$k = \begin{cases} 1 \; ; \; \sigma_m \geq \xi_t \\ k_1(\sigma_m) \; ; \; \xi_t > \sigma_m \geq \xi_c \\ k_0 \; ; \; \xi_c > \sigma_m \geq \xi_k \\ k_2(\sigma_m) \; ; \; \xi_k > \sigma_m \end{cases} \tag{2.17}$$

where,

$$k_1(\sigma_m) = 1 + \frac{(1-k_0)[-\xi_t(-2\cdot \xi_c+\xi_t)-2\cdot \xi_c\cdot \sigma_m+\sigma_m^2]}{(\xi_c-\xi_t)^2} \tag{2.18}$$

$$k_2(\sigma_m) = \frac{k_0(\overline{\xi}-\sigma_m)(\overline{\xi}+\sigma_m-2\cdot \xi_k)}{(\overline{\xi}-\xi_k)} \tag{2.19}$$

$\overline{\xi}$: intersection of the loading surface with the hydrostatic axis (Refer Fig.2.12). It can be expressed as,

$$\overline{\xi} = \frac{A}{1-k_0} \tag{2.20}$$

where A is a constant. The value for $\overline{\xi}$ approaches infinity as k_0 gets nearer to unity, thus modifying the loading surface into an open shape.

From experimental results reported by *Launay* and *Gachon*, it is found that the effects of yielding under a pure hydrostatic pressure only occur under high stress conditions and as such, is not of much importance in the analysis of lowly confined concrete structures such as reinforced concrete slabs. The value of k_2 is considered to be constant throughout the calculations here ($k_2=k_0$). Therefore, the final yield and failure surfaces used in the present study can be illustrated as shown in Fig.2.13 [2.23].

The plastic modulus H_p used in the constitutive equation can be calibrated from the base plastic modulus H_p^b obtained from the uniaxial material compressive test. The base plastic modulus can be defined as the slope of the experimental uniaxial compressive stress- plastic strain curve at a given stress level. But since the calibrated results are only valid within the uniaxial compressive stress state ($\sigma_m \leq f_c'/3$), a modification factor proposed by *Chen* and *Han* [2.5] is introduced here to account for the hydrostatic pressure sensitivity and lode angle (θ) dependance. H_p can then be expressed as,

$$H_p = M(\sigma_m, \theta) \cdot H_p^b \tag{2.21}$$

where,

$$M(\sigma_m, \theta) = \begin{cases} f(\sigma_m, \theta) & \text{; if } 0 < f \leq 1 \\ 1 & \text{; otherwise} \end{cases} \tag{2.22}$$

$$f(\sigma_m, \theta) = -\frac{0.15}{(1.4 - \cos\theta)(\sigma_m + 1/3)(\sigma_m + 2.5)} \tag{2.23}$$

Since the effects of hydrostatic pressure in the analysis of impact or impulsive loading of lowly confined concrete structures, such as reinforced concrete slabs, are of not much importance, $M(\sigma_m, \theta)$ can be considered to be equivalent to unity.

Experimental results for concrete materials indicate that under compressive loadings, inelastic volume contraction occurs at the beginning of yielding and volume dilatation occurs at about 75 to 90% of the ultimate stress. A plastic potential other than the loading function is therefore needed to define the flow rule. For simplicity, a functional form of the *Drucker-Prager* type is used.

The *Drucker-Prager* plastic potential,

$$g = \alpha \cdot I_1 + \sqrt{J_2} \tag{2.24}$$

where α represents a measure of plastic volume dilatation. α can be taken as a function of σ_m but it is assumed that it remains constant for convenience. ($\alpha = 0.07$, according to biaxial test results of *Kupfer* et. al. [2.29])

The flow rule then becomes,

$$d\varepsilon_{ij}^p = d\lambda \cdot \frac{dg}{d\sigma_{ij}} = d\lambda \cdot \left(\alpha \cdot \delta_{ij} + \frac{s_{ij}}{2\sqrt{J_2}} \right) \tag{2.25}$$

2.2.3 CONSTITUTIVE MODEL BASED ON CONCRETE PLASTICITY

From the classical theory of plasticity, the total strain increment can be assumed to be the sum of the elastic ($d\varepsilon^e$) as well as the plastic strain increments ($d\varepsilon^p$). It can be expressed as,

$$d\varepsilon_{ij} = d\varepsilon_{ij}^e + d\varepsilon_{ij}^p \tag{2.26}$$

According to *Hooke's* law, the stress increment is determined by increases in the elastic strain component only, which can be expressed as in the following equation,

$$\begin{aligned} d\sigma_{ij} &= D_{ijkl} \cdot d\varepsilon_{kl}^e \\ &= D_{ijkl} \cdot (d\varepsilon_{kl} - d\varepsilon_{kl}^p) \end{aligned} \tag{2.27}$$

where D_{ijkl} represents the material stiffness matrix.

While plastic flow takes place, the consistency condition must hold,

$$\therefore df = 0 \tag{2.28}$$

The loading function,

$$df = \frac{\partial f}{\partial \sigma_{ij}} \cdot d\sigma_{ij} + \frac{\partial f}{\partial \tau} \cdot \frac{d\tau}{d\varepsilon_p} \cdot d\varepsilon_p = 0$$

(2.29)

$$df = \frac{\partial f}{\partial \sigma_{ij}} \cdot D_{ijkl} \cdot (d\varepsilon_{kl} - d\varepsilon_{kl}^p) + \frac{\partial f}{\partial \tau} \cdot H^p \cdot d\varepsilon_p = 0$$

(2.30)

where,

H^p : plastic modulus

$$H^p = \frac{d\tau}{d\varepsilon_p}$$

(2.31)

τ : effective stress

$d\varepsilon_p$: effective plastic strain

$$d\varepsilon_p = \phi \cdot d\lambda$$

(2.32)

ϕ : scalar function of stress state

$$d\lambda = \frac{1}{h} \cdot \frac{\partial f}{\partial \sigma_{pq}} \cdot D_{pqkl} \cdot d\varepsilon_{kl}$$

(2.33)

$$h = \frac{\partial f}{\partial \sigma_{mn}} \cdot D_{mnpq} \cdot \frac{\partial g}{\partial \sigma_{pq}} - H^p \cdot \frac{\partial f}{\partial \tau} \cdot \phi$$

(2.34)

For the *Drucker-Prager* type of plastic potential,

$$\phi = \frac{\sqrt{2} \cdot (\alpha \cdot I_1 + \sqrt{J_2})}{\sqrt{3} \cdot \rho_c \cdot k}$$

(2.35)

$$\frac{\partial f}{\partial \tau} = -\left(\frac{\sqrt{2}}{\sqrt{3}} + \frac{k}{3} \cdot \frac{d\rho_c}{d\sigma_m} + \frac{\rho_c}{3} \cdot \frac{dk}{d\sigma_m} \right)$$

(2.36)

The total constitutive equation can be expressed in terms of the total stress increment $d\sigma_{ij}$ and the total strain increment $d\varepsilon_{kl}$. The expression for the plastic strain increment can be derived by substituting Eq.(2.33) into Eq.(2.25), bringing about the following relation,

$$d\varepsilon_{ij}^p = \frac{1}{h} \cdot \frac{\partial g}{\partial \sigma_{ij}} \cdot \frac{\partial f}{\partial \sigma_{pq}} \cdot D_{pqkl} \cdot d\varepsilon_{kl}$$

(2.37)

Substituting Eq.(2.37) into Eq.(2.27) leads to the following constitutive equation,

$$d\sigma_{ij} = (D_{ijkl} + D_{ijkl}^p) \cdot d\varepsilon_{kl}$$

(2.38)

where the plastic stiffness tensor has the form,

$$D_{ijkl}^p = -\frac{1}{h}\, H_{ij}^*\cdot H_{kl}$$

(2.39)

$$H_{ij}^* = D_{ijmn}\cdot \frac{\partial g}{\partial \sigma_{mn}}$$

(2.40)

$$H_{kl} = \frac{\partial f}{\partial \sigma_{pq}}\cdot D_{pqkl}$$

(2.41)

The *Drucker-Prager* form shown in Eq.(2.33) is used as the potential function. The differential function for f in the *Ottosen* model can be expressed as,

$$\frac{\partial f}{\partial \sigma_{ij}} = B_0\cdot \delta_{ij} + B_1\cdot s_{ij} + B_2\cdot t_{ij}$$

(2.42)

where,

$$B_0 = \frac{\partial f}{\partial I_1} = -\frac{\partial k}{\partial I_1}\cdot \rho_f - k\cdot A_0$$

(2.43)

$$B_1 = \frac{\partial f}{\partial J_2} = \frac{1}{\rho} - k\cdot A_1$$

(2.44)

$$B_2 = \frac{\partial f}{\partial J_3} = -k\cdot A_2$$

(2.45)

$$A_0 = \frac{\partial \rho_f}{\partial I_1} = \frac{\partial \rho_f}{3\cdot \partial \sigma_m} = -\frac{2b}{h_1}$$

(2.46)

$$A_1 = \frac{\partial \rho_f}{\partial J_2} = \frac{1}{2a}\left(-\sqrt{2} + \frac{2\lambda}{h_1}\right)\cdot \frac{d\lambda}{d\theta}\cdot \frac{\partial \theta}{\partial J_2}$$

(2.47)

$$A_2 = \frac{\partial \rho_f}{\partial J_3} = \frac{1}{2a}\left(-\sqrt{2} + \frac{2\lambda}{h_1}\right)\cdot \frac{d\lambda}{d\theta}\cdot \frac{\partial \theta}{\partial J_3}$$

(2.48)

$$h_1 = \sqrt{2\cdot \lambda^2 - 8a\cdot (3b\cdot \sigma_m - 1)}$$

(2.49)

$$\frac{\partial \theta}{\partial J_2} = \frac{3\sqrt{3}}{4\cdot \sin(3\theta)}\cdot \frac{J_3}{(J_2)^{5/2}}$$

(2.50)

$$\frac{\partial \theta}{\partial J_3} = -\frac{\sqrt{3}}{2\cdot \sin(3\theta)}\cdot \frac{1}{(J_2)^{3/2}}$$

(2.51)

$$\frac{d\lambda}{d\theta} = \begin{vmatrix} \dfrac{-k_1 \cdot k_2 \cdot \sin(3\theta) \cdot \sin[(1/3)\cos^{-1}(k_2 \cdot \cos(3\theta))]}{\sin[\cos^{-1}(k_2 \cdot \cos(3\theta))]} & ; \text{for } \cos(3\theta) \geq 0 \\[4mm] \dfrac{-k_1 \cdot k_2 \cdot \sin(3\theta) \cdot \sin[\pi/3 - (1/3)\cos^{-1}(k_2 \cdot \cos(3\theta))]}{\sin[\cos^{-1}(-k_2 \cdot \cos(3\theta))]} & ; \text{for } \cos(3\theta) < 0 \end{vmatrix}$$

(2.52)

Using the *Drucker-Prager* criterion for the nonassociated flow rule,

$$\frac{\partial g}{\partial \sigma_{ij}} = \alpha \cdot \delta_{ij} + \frac{s_{ij}}{2\sqrt{J_2}}$$

(2.53)

By substituting Eqs.(2.42) and (2.53), the plastic stiffness tensor can then be expressed as,

$$D^p_{ijkl} = -\frac{1}{h} H^*_{ij} \cdot H_{kl}$$

(2.54)

where,

$$h = 2G \cdot \left(3B_0 \cdot \alpha \cdot \frac{1+v}{1-2v} + B_1 \sqrt{J_2} + \frac{3B_2}{2\sqrt{J_2}} \right) - \phi \cdot H^p \cdot \frac{\partial f}{\partial \tau}$$

(2.55)

$$H_{ij} = 2G \cdot \left(B_0 \cdot \frac{1+v}{1-2v} \delta_{ij} + B_1 \cdot s_{ij} + B_2 \cdot t_{ij} \right)$$

(2.56)

$$H^*_{ij} = 2G \cdot \left(\alpha \cdot \frac{1+v}{1-2v} \delta_{ij} + \frac{s_{ij}}{2\sqrt{J_2}} \right)$$

(2.57)

From the equations, it can be seen that the stiffness tensor is not symmetrical. To implement the equations into the finite element method, it is of convenience to express the equations explicitly as,

$$H_{xx} = 2G \cdot \left[B_0 \cdot \frac{1+v}{1-2v} + B_1 \cdot s_{xx} + B_2 \cdot \left(s_{xx}^2 + s_{xy}^2 + s_{xz}^2 - \frac{2J_2}{3} \right) \right]$$

(2.58)

$$H_{yz} = 2G \cdot \left[B_1 \cdot s_{yz} + B_2 \cdot \left(s_{xy} \cdot s_{xz} + s_{yy} \cdot s_{yz} + s_{yz} \cdot s_{zz} \right) \right]$$

(2.59)

and,

$$H^*_{xx} = 2G \cdot \left(\alpha \cdot \frac{1+v}{1-2v} + \frac{s_{xx}}{2\sqrt{J_2}} \right)$$

(2.60)

$$H^*_{yz} = \frac{G}{\sqrt{J_2}} \cdot s_{yz}$$

(2.61)

and etc. The stress-strain relation then becomes,

$$\{d\sigma\} = ([D] + [D]^p) \cdot \{d\varepsilon\}$$

(2.62)

where,

$$\{d\sigma\} = [d\sigma_x, d\sigma_y, d\sigma_z, d\tau_{yz}, d\tau_{xz}, d\tau_{xy}]^T \tag{2.63}$$

$$\{d\varepsilon\} = [d\varepsilon_x, d\varepsilon_y, d\varepsilon_z, d\gamma_{yz}, d\gamma_{xz}, d\gamma_{xy}]^T \tag{2.64}$$

and,

$$[D]^p = -\frac{1}{h} [H^*]\cdot [H]^T \tag{2.65}$$

where,

$$[H] = [H_{xx}, H_{yy}, H_{zz}, H_{yz}, H_{xz}, H_{xy}] \tag{2.66}$$

$$[H^*] = [H^*_{xx}, H^*_{yy}, H^*_{zz}, H^*_{yz}, H^*_{xz}, H^*_{xy}] \tag{2.67}$$

Based on the assumed loading function, the effective stress τ and effective strain increment $d\varepsilon_p$ for a multiaxial stress state can be calibrated to results from a uniaxial material compressive test. The uniaxial compressive stress-plastic strain curve derived can be applied here.

In the uniaxial compressive test, the state of stress can be given by $(-\tau, 0, 0)$, in reference to a triaxial stress condition. Substitution into the loading function brings about,

$$\rho = \frac{\sqrt{2}}{\sqrt{3}}\tau \tag{2.68}$$

$$\sigma_m = -(1/3)\cdot \tau \tag{2.69}$$

$$\rho_f = \rho_c = \frac{1}{2a}\left[-\sqrt{2}\cdot\lambda + \sqrt{2\cdot\lambda^2 + 8a(b\cdot\tau + 1)}\right] \tag{2.70}$$

The loading function then becomes,

$$f = \sqrt{2/3}\cdot\tau - k\cdot\rho_c = 0 \tag{2.71}$$

The effective stress τ can then be defined as,

$$\tau = \frac{\sqrt{3}}{\sqrt{2}} k\cdot\rho_c \tag{2.72}$$

The corresponding effective plastic strain increment $d\varepsilon_p$ can be defined in terms of the plastic work per unit volume,

$$dW_p = \tau\cdot d\varepsilon_p \tag{2.73}$$

On the other hand, plastic work can also be expressed as (Eq.(2.85)),

$$dW_p = \sigma_{ij}\cdot d\varepsilon^p_{ij} = \sigma_{ij}\cdot d\lambda\cdot\frac{\partial g}{\partial\sigma_{ij}} \tag{2.74}$$

From Eqs.(2.73) and (2.74),

$$d\varepsilon_p = \phi \cdot d\lambda \tag{2.75}$$

where,

$$\phi = \frac{1}{\tau} \cdot \frac{\partial g}{\partial \sigma_{ij}} \cdot \sigma_{ij} \tag{2.76}$$

For the nonassociated flow rule based on the *Drucker-Prager* criterion,

$$\phi = \frac{\sqrt{2} \cdot (\alpha \cdot I_1 + \sqrt{J_2})}{\sqrt{3} \cdot \rho_c \cdot k} \tag{2.77}$$

The consistency equation can also be expressed as,

$$df = \frac{\partial f}{\partial \sigma_{ij}} \cdot d\sigma_{ij} + \frac{\partial f}{\partial \tau} \cdot d\tau \tag{2.78}$$

Noting that $\tau = -\sigma_{33}$, $\rho = -\sqrt{2/3} \cdot \sigma_{33}$ and $\sigma_m = (1/3) \cdot \sigma_{33}$,
Differentiating Eq.(2.71) results in,

$$\frac{\partial f}{\partial \tau} = -\left(\frac{\sqrt{2}}{\sqrt{3}} + \frac{k}{3} \cdot \frac{d\rho_c}{d\sigma_m} + \frac{\rho_c}{3} \cdot \frac{dk}{d\sigma_m} \right) \tag{2.79}$$

The gradient of the incremental stress-strain relation in Eq.(2.62) can be written as,

$$\left(\frac{d\sigma}{d\varepsilon} \right) = [D] + [D]^p = [D] - \frac{1}{h} \cdot [H^*] \cdot [H]^{\mathrm{T}} \tag{2.80}$$

For the uniaxial compressive test state,

$$\frac{d\sigma}{d\varepsilon} = D_{xx} - \frac{1}{h} \cdot H_{xx}^* \cdot H_{xx} \tag{2.81}$$

$$\therefore \frac{d\sigma}{d\varepsilon} = D_{xx} - \frac{D_{xx} \cdot (\partial g/\partial \sigma_{xx}) \cdot (\partial f/\partial \sigma_{xx}) \cdot D_{xx}}{(\partial f/\partial \sigma_{xx}) \cdot D_{xx} \cdot (\partial g/\partial \sigma_{xx}) - H^p \cdot (\partial f/\partial \tau) \cdot \phi} \tag{2.82}$$

The gradient $d\sigma/d\varepsilon$ can be denoted by $\beta \cdot \dot{D}_{xx} (= \beta \cdot E_x)$ where β shows the amount of decrease in initial stiffness $D_{xx} (= E_x)$, bringing about the following equation,

$$\beta \cdot D_{xx} = D_{xx} - \frac{D_{xx} \cdot (\partial g/\partial \sigma_{xx}) \cdot (\partial f/\partial \sigma_{xx}) \cdot D_{xx}}{(\partial f/\partial \sigma_{xx}) \cdot D_{xx} \cdot (\partial g/\partial \sigma_{xx}) - H^p \cdot (\partial f/\partial \tau) \cdot \phi} \tag{2.83}$$

Solving the equation,

$$-H^p \cdot \frac{\partial f}{\partial \tau} \cdot \phi = -\frac{\beta \cdot (\partial f/\partial \sigma_{xx}) \cdot (\partial g/\partial \sigma_{xx}) \cdot D_{xx}}{1 - \beta} \tag{2.84}$$

Fig.2.14 shows an example of the uniaxial stress-strain and stress-plastic strain relation obtained from tests on concrete cylinders and the corresponding effective stress-effective strain relation, while Fig.2.15 shows the relation between the uniaxial compressive test results and the calibrated values for use in the multiaxial stress state.

2.2.4 MATERIAL CHARACTERISTICS FOR VARIOUS CONCRETE MATERIALS

In order to study the effects of different concrete material characteristics on impact performance of concrete structures, four different types of concrete materials are mainly applied in this study. The types of concrete are normal strength concrete (RC), high strength concrete (HRC), lightweight aggregate concrete (LRC) and steel fiber reinforced concrete (SFRC). A typical example of the stress-strain relations for the four types of concrete are shown in Figs.2.16 and 2.17[2.30] while a list of the corresponding mechanical properties are given in Table 2.3.

The uniaxial tensile stress-strain relation of steel fiber reinforced concrete is a function of the amount of steel fibers in the concrete matrix material. By applying the method proposed by *Hannant* [2.31], the critical volume of steel fibers is estimated at 1.33% and the corresponding tensile uniaxial stress-strain relation is as shown in Fig.2.17(a). As a comparison, the uniaxial stress-strain relation for a volume of fiber of 3% is also calculated and is given in Fig.2.17(b). The tensile yield strain of the steel fiber reinforced concrete is similar to that of normal strength concrete, but after formation of cracks, the steel fibers would then be effective in bridging the cracks and allow the stresses to flow between cracks. It is assumed here that there are no significant changes in the uniaxial stress-strain relation under compression between normal strength concrete and steel fiber reinforced concrete.

2.3 FAILURE AND YIELD MODEL FOR REINFORCEMENT

The failure and yield model for reinforcement is similar to the method proposed by *Isobata* [2.32]. It is assumed here that the uniaxial stress-strain relation is similar in both tension and compression. Two different types of reinforcement are considered throughout this study, i.e., steel based reinforcement and fiber reinforced plastic (FRP) based reinforcement.

2.3.1 STEEL REINFORCEMENT

Mainly two different types of reinforcement are employed in the analyses. They are the normal reinforcement bar (SD35, JIS specification) with a yield stress of $3500kgf/cm^2$ and the high tensile strength reinforcement bar (SD70, JIS specification) with a yield stress of $7000kgf/cm^2$. The idealized uniaxial tensile stress-strain relations are shown in Fig.2.18 while the corresponding mechanical properties are given in Table 2.4.

2.3.2 FRP REINFORCEMENT [2.33, 2.34]

FRP reinforcing bars are made of a combination of long fiber strands being fabricated by different types of synthetic resin and therefore, there is no fix standards for these FRP reinforcing bars, in comparison with structural steel. In order to set a certain standard criterion for different types of bars, the energy absorption capacity, denoted by the area enclosed in the stress-strain curve, is selected. Eight types of FRP reinforcement (Type A1 ~ D) and two types of steel reinforcement (SD35, SD70) are considered. The stress-strain relations and the mechanical properties are as shown in Fig.2.18 and Table 2.4, respectively. The energy absorption capacity for all reinforcements, with the exception of the D reinforcements, are set equivalent to the SD35 steel reinforcement. At present, most FRP reinforcements do not have a yielding point or a plastic region, and fail directly after exceeding the elastic region, thus causing brittle type of failure to be more likely to occur. However, it is possible to duplicate a yielding phenomenon by combining two or more types of fibers in a single reinforcing bar, and the D type of reinforcement shows a real example.

The "A" and "C" series of FRP reinforcements are set with an elastic modulus of 1.3 times and 0.5 times that of structural steel, respectively, while the "B" reinforcement has an equivalent elastic modulus to structural steel. The A1, B and C1 reinforcements have a fully elastic regime till failure, while the A2 and C2 reinforcements reach a plastic stage at 70% the failure load of the A1 and C1 reinforcements, respectively. On the other hand, the A3 and C3 reinforcements are set in such a way that the ultimate tensile strain reaches 45000μ, which is relatively similar to structural steel.

The uniaxial stress-strain relation during compression is assumed to be similar to that of tension. But in reality, the compressive strength can be expected to be lower because the compressive strength would be mainly due to the compressive strength of the fabricating synthetic resin material. But since there is no data on the compressive behavior of FRP based materials at present, it is assumed to be similar during compression.

2.4 CRITERIA OF LOADING AND UNLOADING

The stress space is assumed to be based on *Drucker's* stability postulate. Applying the principle of virtual work, the work done in a work cycle, dW, which is non-negative, can be expressed as (stability in cycle in small - Fig.2.19(a)),

$$dW = \oint d\sigma_{ij} \cdot d\varepsilon_{ij}^p \geq 0$$

(2.85)

where $\int$ is the integral taken over a cycle of applying and removing the added stress set $d\sigma_{ij}$ and the plastic strain increment defined by Eq.(2.25). The equal sign is valid only when no plastic strain occurred in the cycle, i.e., in the elastic stages.

The incremental elasto-plastic stress-strain relation is valid only during plastic loading. Therefore, the stress condition has to be verified prior to applying the constitutive equation. The elastic stress-strain equation has to be applied during elastic loading.

The definitions for the loading condition in stress space can be given as,

$$f = 0 \text{ and } \frac{\partial f}{\partial \sigma_{ij}} \cdot d\sigma_{ij} > 0 \rightarrow \text{loading} \qquad ; d\sigma_{ij}^p \neq 0$$

(2.86)

$$f = 0 \text{ and } \frac{\partial f}{\partial \sigma_{ij}} \cdot d\sigma_{ij} = 0 \rightarrow \text{neutral loading} \quad ; d\sigma_{ij}^p = 0$$

(2.87)

$$f = 0 \text{ and } \frac{\partial f}{\partial \sigma_{ij}} \cdot d\sigma_{ij} < 0 \rightarrow \text{unloading} \qquad ; d\sigma_{ij}^p = 0$$

(2.88)

Based on Fig.2.19(b), under loading, the stress condition is such that the stress state moves outward from one loading surface to another new surface, while under neutral loading the stress point moves along the particular loading surface. The stress state is assumed to move inwards according to the former loading condition to a previous loading surface during unloading. Only the elastic strain components will decrease during unloading.

2.5 DYNAMIC ANALYSIS [2.35 - 2.37]

2.5.1 INCREMENTAL EQUATION OF MOTION

Since the time in which impact or impulsive loading takes place is comparatively short (transient phenomena), generally, the effects of viscous damping becomes small enough to be ignored in the case of concrete structures. But as a general case, the effects of viscous damping will be included in the equations here. The *Newmark*-β method, which is an implicit method, will be employed here.

The semidiscrete equation of motion at the nodes of a finite assembly can be treated as,

$$[M] \cdot \{\ddot{U}\}_t + [C] \cdot \{\dot{U}\}_t + [K] \cdot \{U\}_t = \{R\}_t \tag{2.89}$$

where, $[M]$, $[C]$ and $[K]$ represent the mass, viscous damping and stiffness matrices respectively, while $\{R\}_t$ is the external load vector. $\{\ddot{U}\}_t$, $\{\dot{U}\}_t$ and $\{U\}_t$ are the acceleration, velocity and displacement vectors for the finite element assembly, respectively. The subscript "t" is used for quantities at time t and a dot denotes a derivative with respect to time.

The response history is divided into discrete time increments, Δt, which are of equal length. The system is calculated for each time increment with properties determined at the beginning of the interval. The discretized equation for Eq.(2.89) during a discrete time increment of Δt is shown in the following equation ($[C]$ is assumed as being independent to t),

$$[M] \cdot \{\Delta \ddot{U}\}_{t \to t+\Delta t} + [C] \cdot \{\Delta \dot{U}\}_{t \to t+\Delta t} + [K] \cdot \{\Delta U\}_{t \to t+\Delta t} = \{\Delta R\}_{t \to t+\Delta t} \tag{2.90}$$

As Eq.(2.90) is simply an approximative equation of motion, it is solved using the *Newmark-β* method which consists of the following equations,

$$\{U\}_{t+\Delta t} = \{U\}_t + \Delta t \cdot \{\dot{U}\}_t + \Delta t^2 \cdot [(1/2 - \beta)\{\ddot{U}\}_t + \beta\{\ddot{U}\}_{t+\Delta t}] \tag{2.91}$$

$$\{\dot{U}\}_{t+\Delta t} = \{\dot{U}\}_t + (1/2) \cdot \Delta t \cdot (\{\ddot{U}\}_t + \{\ddot{U}\}_{t+\Delta t}) \tag{2.92}$$

The discretized equation for acceleration can be obtained from Eqs.(2.91) and (2.92) which describe the evolution of the approximative solution as follows,

$$\{\Delta \ddot{U}\}_{t \to t+\Delta t} = \frac{1}{\beta \cdot \Delta t^2} \{\Delta U\}_{t \to t+\Delta t} - \frac{1}{\beta \cdot \Delta t} \{\dot{U}\}_t - \frac{1}{2 \cdot \beta} \{\ddot{U}\}_t \tag{2.93}$$

The discretized equation for velocity is obtained from Eq.(2.92) as follows,

$$\{\Delta \dot{U}\}_{t \to t+\Delta t} = \frac{1}{2} \Delta t \cdot [\{\ddot{U}\}_t + \{\ddot{U}\}_{t+\Delta t}] \tag{2.94}$$

The following equation can be derived from Eq.(2.91),

$$\{\ddot{U}\}_{t+\Delta t} = \frac{1}{\beta \cdot \Delta t^2} \{\Delta U\}_{t \to t+\Delta t} - \frac{1}{\beta \cdot \Delta t} \{\dot{U}\}_t - \left(\frac{1}{2\beta} - 1\right) \{\ddot{U}\}_t \tag{2.95}$$

By substituting Eq.(2.95) into Eq.(2.94), the following equation is the obtained,

$$\{\Delta \dot{U}\}_{t \to t+\Delta t} = \frac{1}{2\beta \cdot \Delta t} \{\Delta U\}_{t \to t+\Delta t} - \frac{1}{2\beta} \{\dot{U}\}_t + \frac{(4\beta - 1)}{4\beta} \Delta t \cdot \{\ddot{U}\}_t \tag{2.96}$$

Substituting Eqs.(2.93) and (2.96) into Eq.(2.90) gives the following equation,

$$\left([K] + \frac{1}{2\beta \cdot \Delta t} [C] + \frac{1}{\beta \cdot \Delta t^2} [M]\right) \cdot \{\Delta U\}_{t \to t+\Delta t} = \{\Delta R\}_{t \to t+\Delta t} +$$

$$[C] \cdot \left(\frac{1}{2\beta} \{\dot{U}\}_t - \frac{(4\beta - 1)}{4\beta} \Delta t \cdot \{\ddot{U}\}_t\right) + [M] \cdot \left(\frac{1}{\beta \cdot \Delta t} \{\dot{U}\}_t + \frac{1}{2\beta} \{\ddot{U}\}_t\right) \tag{2.97}$$

In the above equations, the parameter β and the discrete time increment Δt are closely related to the accuracy of the integration and also the stability of the dynamic solution. In this study, the parameter of $\beta=1/4$ is used because it satisfies the necessary stability conditions. The properties of members of the *Newmark*-β method are listed in Table 2.5[2.38]. Based on a few trial calculations, the most appropriate and stable values for the time increment are selected (see Table 2.6). The integration of the equation of motion (Eq.(2.90)) with respect to time can be obtained by solving Eq.(2.97). In order to improve the accuracy, an iterative method (Refer following Section) is included in the following form,

$$[M] \cdot \left(\frac{1}{\beta \cdot \Delta t^2} \cdot \{\Delta U\}^i_{t \rightarrow t+\Delta t} - \frac{1}{\beta \cdot \Delta t} \cdot \{\dot{U}\}_t - \frac{1}{2\beta} \cdot \{\ddot{U}\}_t \right) +$$

$$[C] \cdot \left(\frac{1}{2\beta \cdot \Delta t} \cdot \{\Delta U\}^i_{t \rightarrow t+\Delta t} - \frac{1}{2\beta} \cdot \{\dot{U}\}_t + \frac{(4\beta-1)}{4\beta} \cdot \Delta t \cdot \{\ddot{U}\}_t \right) +$$

$$[K]_{t+\Delta t} \cdot \{\Delta U\}^i_{t \rightarrow t+\Delta t} =$$

$$\{\Delta R\}_{t \rightarrow t+\Delta t} - [K]_{t+\Delta t} \cdot \{\Delta U\}^{i-1}_{t \rightarrow t+\Delta t} -$$

$$\frac{1}{2\beta \cdot \Delta t} \cdot [C] \cdot \{\Delta U\}^{i-1}_{t \rightarrow t+\Delta t} - \frac{1}{\beta \cdot \Delta t^2} \cdot [M] \cdot \{\Delta U\}^{i-1}_{t \rightarrow t+\Delta t}$$

$$(2.98)$$

where, $\{\Delta U\}^i_{t \rightarrow t+\Delta t}$ is the increase in displacement vector during the iterative count of i.

2.5.2 ITERATIVE METHOD FOR SOLUTION OF EQUATION OF MOTION

The flow of the iterative method (step-by-step integration) applied in the analysis can be summarized as:

[1] Initialization

(1)Formation of the effective stiffness matrix $[K^*]$,

$$[K^*] = [K] + \frac{1}{2\beta \cdot \Delta t} \cdot [C] + \frac{1}{\beta \cdot \Delta t^2} \cdot [M]$$

$$(2.99)$$

[2] For each time step:

(1) Calculation of the constant parts of the effective load vector,

$$\{\Delta R^*\}_{t \rightarrow t+\Delta t} = \{\Delta R\}_{t \rightarrow t+\Delta t} + [C] \cdot \left(\frac{1}{2\beta} \cdot \{\dot{U}\}_t - \frac{(4\beta-1)}{4\beta} \cdot \Delta t \cdot \{\ddot{U}\}_t \right)$$

$$+ [M] \cdot \left(\frac{1}{\beta \cdot \Delta t} \cdot \{\dot{U}\}_t + \frac{1}{2\beta} \cdot \{\ddot{U}\}_t \right)$$

$$(2.100)$$

(2) Initialization
   Iteration step $i=0$
   Variable parts of the effective load vector,
   $i=0$

$$\{R_{Err}\}^{i=0} = 0$$

$$(2.101)$$

(3) Iteration
(a) $i=i+1$
(b) Formation of effective load vector,

$$\{\Delta R^*\}^i_{t\to t+\Delta t} = \{\Delta R^*\}_{t\to t+\Delta t} + \{R_{Err}\}^{i-1}$$

(2.102)

(c) Calculation of discrete increment in displacement,

$$\{\Delta U\}^i_{t\to t+\Delta t}$$

using the band matrix method [2.39].

(d) Change of initial load vector caused by nonlinear behavior of material,

$$\{\Delta R_{Err}\}^i = -[K]_{t+\Delta t} \cdot \{\Delta U\}^i_{t\to t+\Delta t} - \frac{1}{2\beta \cdot \Delta t}\, [C] \cdot \{\Delta U\}^i_{t\to t+\Delta t}$$

$$- \frac{1}{\beta \cdot \Delta t^2}\, [M] \cdot \{\Delta U\}^i_{t\to t+\Delta t}$$

(2.103)

$$\{R_{Err}\}^i = \{R_{Err}\}^{i-1} + \{\Delta R_{Err}\}^i$$

(2.104)

(e) Iteration convergence,

$$\left| \{\Delta R_{Err}\}^i \right| / \left| \{R_{Err}\}^i \right| \le tol.$$

(2.105)

[3] Calculation of displacement, acceleration and velocity,

$$\{U\}_{t+\Delta t} = \{U\}_t + \{\Delta U\}^i_{t\to t+\Delta t}$$

(2.106)

$$\{\ddot{U}\}_{t+\Delta t} = \{\ddot{U}\}_t + \frac{1}{\beta \cdot \Delta t^2}\, \{\Delta U\}^i_{t\to t+\Delta t} - \frac{1}{\beta \cdot \Delta t}\, \{\dot{U}\}_t - \frac{1}{2\beta}\, \{\ddot{U}\}_t$$

(2.107)

$$\{\dot{U}\}_{t+\Delta t} = \{\dot{U}\}_t + \frac{1}{2}\, \Delta t \cdot (\{\ddot{U}\}_t + \{\ddot{U}\}_{t+\Delta t})$$

(2.108)

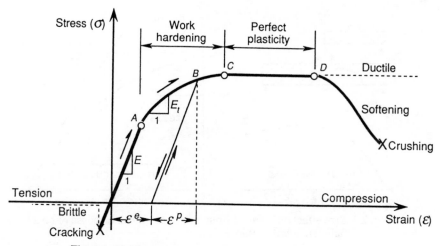

Fig. 2.1 Uniaxial stress-strain curve for concrete

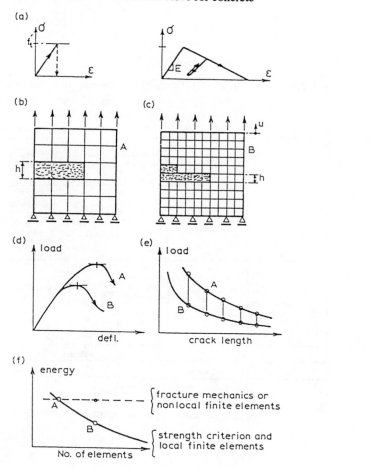

Fig. 2.2 Spurious mesh sensitivity

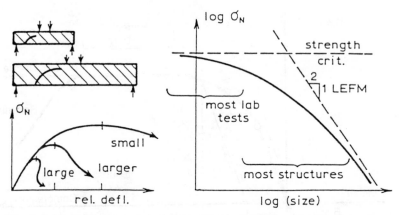

Fig. 2.3 Size effect on nominal strength and post-peak response of structures failing in a brittle manner

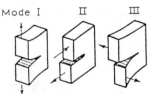

Fig. 2.4 Modes I, II and III (Opening, plane shear, anti-plane shear fractures)

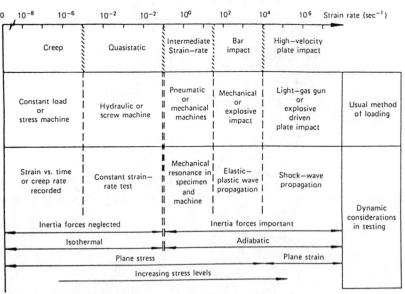

10⁶	10⁴	10²	10⁰	10⁻²	10⁻⁴	10⁻⁶	10⁻⁸

Fig. 2.5 Dynamic aspects of mechanical testing

Table 2.1 Governing equations and considerations for different loading conditions

LOAD		DYNAMIC CONSIDERATIONS	GOVERNING EQUATIONS
Static			Eq. of equilibrium $F = K \cdot x$
Dynamic	Soft impact (Low strain rate)	Inertia forces	Eq. of equilibrium $F = M \cdot \ddot{x} + C \cdot \dot{x} + K \cdot x$
	Hard impact (High strain rate)	Inertia forces + Thermodynamics	Eq. of equilibrium $F = M \cdot \ddot{x} + C \cdot \dot{x} + K \cdot x$ + Eq. of state

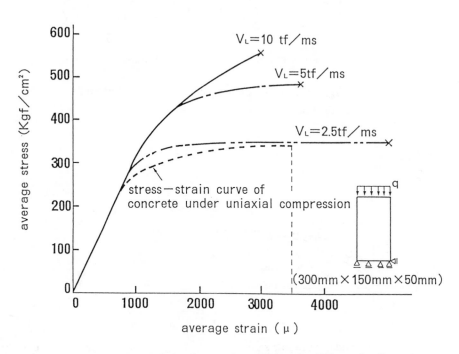

Fig. 2.6 Stress-strain relation for concrete cube under different loading rates

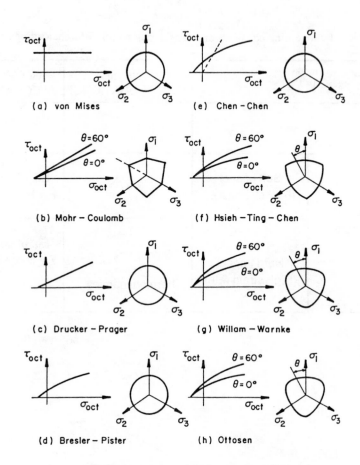

Fig. 2.7 Triaxial failure models for concrete materials

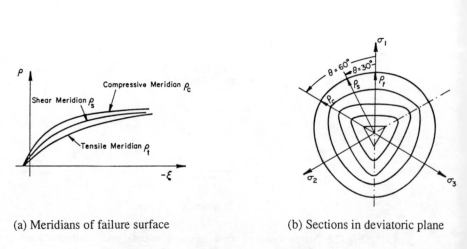

(a) Meridians of failure surface

(b) Sections in deviatoric plane

Fig. 2.8 Basic features of a failure surface

Table 2.2 Parameter values for *Ottosen's* model

$\overline{f_t}$	a	b	k_1	k_2
0.08	1.8076	4.0962	14.4863	0.9914
0.10	1.2759	3.1962	11.7365	0.9801
0.12	0.9218	2.5969	9.9110	0.9647

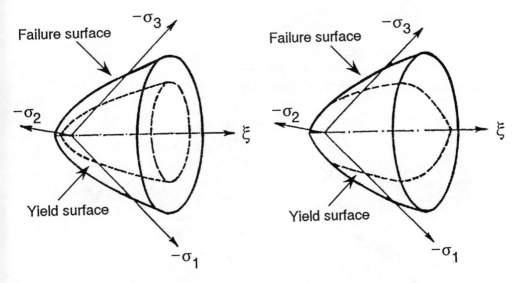

(a) Classical plasticity model (b) Plasticity model by *Chen & Han*

Fig. 2.9 Yield and failure surfaces

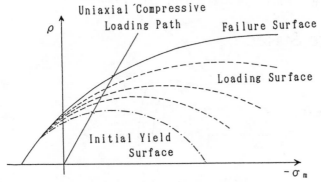

Fig. 2.10 Nonuniform hardening plasticity model (by *Chen & Han*)

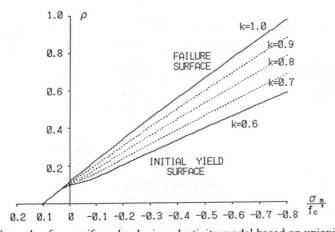

Fig. 2.11 Example of nonuniform hardening plasticity model based on uniaxial compressive test

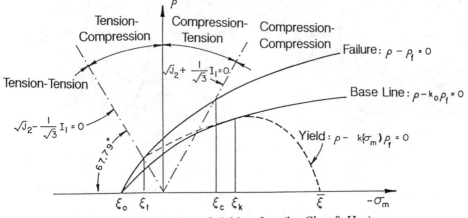

Fig. 2.12 Construction of yield surface (by *Chen & Han*)

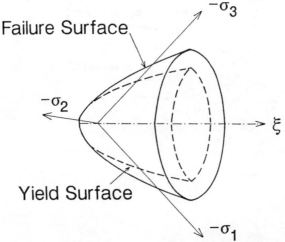

Fig. 2.13 Yield and failure surfaces adopted in this study

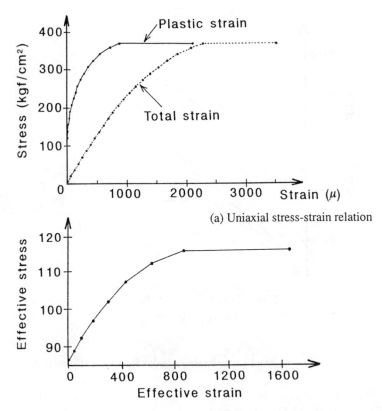

(a) Uniaxial stress-strain relation

(b) Effective stress-effective strain relation

Fig. 2.14 Example of stress-strain relation from uniaxial compressive test

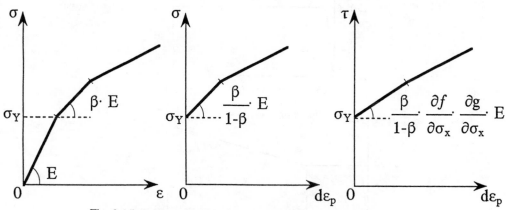

Fig. 2.15 Calibrated effective stress-effective strain relation

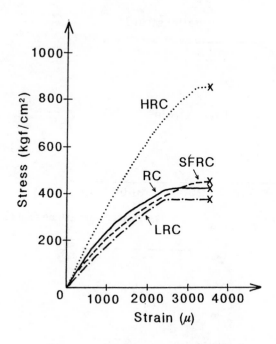

Fig. 2.16 Typical compressive uniaxial stress-strain relations for concrete materials

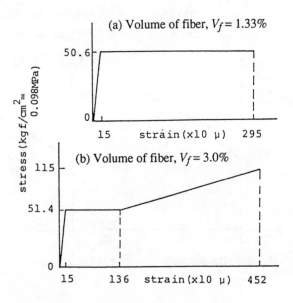

Fig. 2.17 Idealized tensile uniaxial stress-strain relations for steel fiber reinforced concrete

Table 2.3 Mechanical properties of concrete materials

Concrete type	Uniaxial compressive strength (kgf/cm^2)	Tensile yield strain (μ)	Tensile rupture strain (μ)
Normal strength	350 ~ 450	150	150
High strength	850 ~ 1050	250	250
Lightweight aggregate	300 ~ 400	150	150
Steel fiber (Vf = 1.33%)	350 ~ 450	150	2950
Steel fiber (Vf = 3.0%)	350 ~ 450	150	4520

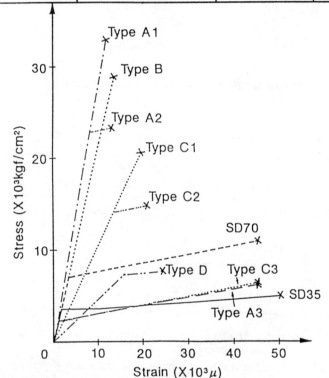

Fig. 2.18 Idealized tensile uniaxial stress-strain relations for reinforcement

Table 2.4 Mechanical properties of reinforcement

Reinforcement	Yield stress (kgf/cm^2)	Yield strain (μ)	Tensile strength (kgf/cm^2)	Tensile strain (μ)	Young's modulus (x10^{6}kgf/cm^2) [ratio]	
Steel (SD35)	3500	1667	4800	50000	2.10	[1.0]
Steel (SD70)	7000	3333	10800	45000	2.10	[1.0]
FRP Type A1	-	-	32900	12000	2.73	[1.3]
FRP Type A2	23030	8400	23400	12800	2.73	[1.3]
FRP Type A3	2460	899	6480	45000	2.73	[1.3]
FRP Type B	-	-	28800	13700	2.10	[1.0]
FRP Type C1	-	-	20400	19400	1.05	[0.5]
FRP Type C2	14280	13580	14900	20500	1.05	[0.5]
FRP Type C3	2650	2517	6520	45000	1.05	[0.5]
FRP Type D	7090	15720	7490	23700	0.451	[0.21]

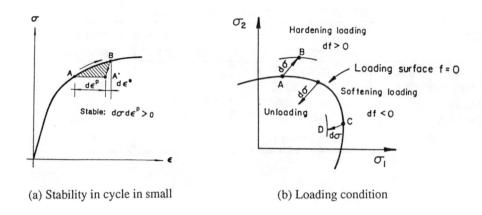

(a) Stability in cycle in small   (b) Loading condition

Fig. 2.19 *Drucker-Prager's* postulate in stress space

Table 2.5 Properties of members of the *Newmark* family of methods

Method	Type	β	γ	Stability condition[2]	Order of accuracy[3]
Average acceleration (trapezoidal rule)	Implicit	$\frac{1}{4}$	$\frac{1}{2}$	Unconditional	2
Linear acceleration	Implicit	$\frac{1}{6}$	$\frac{1}{2}$	$\Omega_{crit} = 2\sqrt{3} \cong 3.464$	2
Fox-Goodwin (royal road)	Implicit	$\frac{1}{12}$	$\frac{1}{2}$	$\Omega_{crit} = \sqrt{6} \cong 2.449$	2
Central difference	Explicit[1]	0	$\frac{1}{2}$	$\Omega_{crit} = 2$	2

Notes: 1. Strictly speaking, M and C need to be diagonal for the central-difference method to be explicit.
2. Stability is based upon the undamped case, in which $\xi = 0$.
3. Second-order accuracy is achieved if and only if $\gamma = \frac{1}{2}$.

Table 2.6 Discrete time increments for analysis of concrete structures

Structure	Type of analysis	Elastic (msec)	Failure (msec)
Beam	Before max. load	0.05	0.0125
	After max. load	0.05	0.1
Slab	Before max. load	0.2	0.05
	After max. load	0.2	0.2
Handrail	Before max. load	0.1	0.025
	After max. load	0.1	0.1

3. NONLINEAR DYNAMIC RESPONSE OF CONCRETE SLAB STRUCTURES SUBJECTED TO SOFT IMPACT LOADS

A 2-dimensional step-by-step finite element method is developed for analysis of reinforced concrete slabs subjected to soft impact. Reinforced concrete slab structures are modeled using the nonlinear dynamic layered finite element method [3.1 - 3.4]. Compared to the conventional 3-dimensional finite element model, the layered finite element method is a quasi-3-dimensional method and thus does not require as much calculation cost and also requires a much smaller amount of memory workload. The analytical procedure is verified by performing tests on full scale reinforced concrete slabs and also guardrails (handrails). The results from the tests, such as deflection and acceleration responses, are then compared with the analytical predictions to check the validity of the model.

Parametric studies, mainly on the effects of loading rates, are performed on reinforced concrete slabs to define the types of failure modes during soft impact loading, i.e., bending failure and punching shear failure. Based on the deformation response and crack propagations from the analytical results, the failure modes are then analytically defined. Next, the dynamic response of concrete slabs are discussed based on effects of transverse shear stresses, effects of unloading during impact loadings, effects of loading rate on various ultimate states, effects of concrete type on the ultimate responses, and the effects of reinforcement on the ultimate behaviors.

3.1 LAYERED FINITE ELEMENT MODEL

A simplified flow of the whole analytical process is shown in Fig.3.1. The analysis can be applied to reinforced concrete slab structures, such as reinforced concrete deck slabs, reinforced concrete barriers or handrails, etc.

3.1.1 FINITE ELEMENT

The finite element model employed here is the 4-node rectangular element. The element is assumed to be subjected to both inplane and bending forces. Provided that the deformations are small, the deformations caused by the two types of forces can be assumed to be independent of each other. But since the analysis undertaken here is to study the ultimate behaviors of concrete slab structures, local deformations can be expected to be quite large and consequently, the coupling effects between the two types of forces may not be of a negligible quantity.

Each element nodal-point has 5 degrees of freedom, i.e., the inplane displacements (u, v), transverse displacement (w) and the sectional rotations around the x and y axes (θ_x, θ_y). The finite element can be considered to be made up of three different parts; the membrane element, the plate-bending element and coupling effects between the two elements. A schematic illustration of the finite element is shown in Fig.3.2.

There is a present trend nowadays of using higher-order elements for finite element analysis. This can be attributed to the development of speedier computers that can handle lots of memory work and also to the fact that the higher-order elements are capable of producing more accurate predictions. The element applied here is kept to the simple 4-node rectangular element because not much improvement can be expected in the analysis of impact or impulsive loadings using the higher-order elements as the phenomena during impact collisions are relatively complex. The increase in workload due to a higher-order element is thus considered not to be cost efficient in this particular case.

The membrane element[3.5, 3.6] applied here is as shown in Fig.3.3. The nodes are taken as $i, j, k,$ and l in the manner shown in the figure. The element is considered to be in a 2-dimensional plane stress field. The nodal displacement vector for an arbitrary node i can be taken as,

$$\{\delta_i\} = \{u_i, v_i\}^T$$

(3.1)

The total displacement vector for an element would then be,

$$\{\delta\} = \{\delta_i, \delta_j, \delta_k, \delta_l\}^T$$

(3.2)

The displacement field of the element can be expressed as,

$$u = \alpha_1 + \alpha_2 x + \alpha_3 y + \alpha_4 xy \tag{3.3}$$

$$v = \beta_1 + \beta_2 x + \beta_3 y + \beta_4 xy \tag{3.4}$$

where α_1, α_2, α_3, α_4, β_1, β_2, β_3, and β_4 are 8 unknowns for each element.

Substituting the coordinates of the 4 nodes and solving for the 8 unknowns in the above equations bring about the following expressions,

$$u = \frac{1}{ab}[(ab-bx-ay+xy)u_i + (bx-xy)u_k + xyu_l + (ay-xy)u_j] \tag{3.5}$$

$$v = \frac{1}{ab}[(ab-bx-ay+xy)v_i + (bx-xy)v_k + xyv_l + (ay-xy)v_j] \tag{3.6}$$

Eqs.(3.5) and (3.6) can also be expressed as,

$$\left\{ \begin{array}{c} u \\ v \end{array} \right\} = [N]\{\delta\} = [\ [I]N_i\ [I]N_j\ [I]N_k\ [I]N_l\ ]\{\delta\} \tag{3.7}$$

where $[I]$ is a 2 x 2 unit matrix and N_i, N_j, N_k, and N_l are defined by,

$$\left. \begin{array}{l} N_i = ab - bx - ay + xy \\ N_j = ay - xy \\ N_k = bx - xy \\ N_l = xy \end{array} \right\} \tag{3.8}$$

For plane stress problems, the strains at any arbitrary location within the element can be expressed as,

$$\{\varepsilon_0\} = \left\{ \begin{array}{c} \varepsilon_{0x} \\ \varepsilon_{0y} \\ \varepsilon_{0xy} \end{array} \right\} = \left[\begin{array}{cccccccc} \frac{\partial N_i}{\partial x} & 0 & \frac{\partial N_j}{\partial x} & 0 & \frac{\partial N_k}{\partial x} & 0 & \frac{\partial N_l}{\partial x} & 0 \\ 0 & \frac{\partial N_i}{\partial y} & 0 & \frac{\partial N_j}{\partial y} & 0 & \frac{\partial N_k}{\partial y} & 0 & \frac{\partial N_l}{\partial y} \\ \frac{\partial N_i}{\partial y} & \frac{\partial N_i}{\partial x} & \frac{\partial N_j}{\partial y} & \frac{\partial N_j}{\partial x} & \frac{\partial N_k}{\partial y} & \frac{\partial N_k}{\partial x} & \frac{\partial N_l}{\partial y} & \frac{\partial N_l}{\partial x} \end{array} \right] \left\{ \begin{array}{c} u_i \\ v_i \\ \vdots \\ u_l \\ v_l \end{array} \right\}$$

$$= \left[\begin{array}{cccccccc} -(b-y) & 0 & -y & 0 & b-y & 0 & y & 0 \\ 0 & -(a-x) & 0 & a-x & 0 & -x & 0 & x \\ -(a-x) & -(b-y) & a-x & -y & -x & b-y & x & y \end{array} \right] \{\delta\} \tag{3.9}$$

$$= [B_p]\cdot \{\delta\} \tag{3.10}$$

The stiffness matrix can then be expressed as,

$$[k_p] = \int_v [B_p]^{T.}\ [D]\cdot [B_p]\cdot dV \tag{3.11}$$

The consistent mass matrix is used in the model here.
The finite element for plate-bending is the 4-node rectangular non-conforming plate-bending

element[3.6 - 3.9]. The element is as illustrated in Fig.3.4. At each node, displacements $\{\delta_n\}$ made up of three components are introduced. The three components are namely the displacement in the direction of z axis, w_n, sectional rotation about the x axis, $(\theta_x)_n$, and sectional rotation around the y axis, $(\theta_y)_n$.

The main assumptions for the plate bending theory are:

(1) The domain Ω is of the following special form,

$$\Omega = \{(x,y,z) \in R^3 | z \in [+h/2,-h/2], (x,y) \in A \subset R^2\}$$

(3.12)

where h is the plate thickness and A is its area.

(2) The transverse stress, $\sigma_z = 0$

(3.13)

This assumption brings about the plane stress hypothesis. This assumption mathematically contradicts the fourth assumption but it is not known to cause any major problems. The practical engineering applications of this assumption outweighs the mathematical contradiction caused.

(3) $u_\alpha(x,y,z) = -z \cdot \theta_\alpha(x,y)$

(3.14)

This assumption implies that plane sections remain plane after deformations. θ_α is the rotation of a fiber initially normal to the plate midsurface. Fig.3.5 shows a diagrammatic illustration of the rotation.

(4) $w(x,y,z) = w(x,y)$

(3.15)

By this assumption, it is understood that the transverse displacement, w, remains constant through the thickness of the plate.

The assumptions above refer to what is known as the *Reissner-Mindlin* plate theory. This theory is different from the classical *Kirchhoff* theory of plates, where the transverse shear strains are assumed to be zero. In the *Reissner-Mindlin* theory, transverse shear strains can be taken into consideration by means of the term γ_α shown in Fig.3.5. According to this theory, the plate displacements can be expressed as follows,

$$u = z \cdot \theta_x, \ v = z \cdot \theta_y, \ w = w \ (x,y)$$

(3.16)

The nodal displacement vector for an arbitrary node i can be expressed as,

$$\{\delta_i\} = \{w_i, \theta_{xi}, \theta_{yi}\}^T$$
$$= \{w_i, -(\partial w/\partial y)_i, -(\partial w/\partial x)_i\}^T$$

(3.17)

The element total displacement (plate bending only) can then be expressed by the displacement vectors for the four nodal points,

$$\{\delta^e\} = \{\delta_i, \delta_j, \delta_k, \delta_l\}^T$$

(3.18)

Since the total degree-of-freedom for each element is 12, the shape function, which can be considered as w, can be defined as a 12 degree polynomial function. It is defined as a fourth-order polynomial expression here,

$$w = \alpha_1 + \alpha_2 x + \alpha_3 y + \alpha_4 x^2 + \alpha_5 xy + \alpha_6 y^2 + \alpha_7 x^3 + \alpha_8 x^2 y + \alpha_9 xy^2$$
$$+ \alpha_{10} y^3 + \alpha_{11} x^3 y + \alpha_{12} xy^3 \tag{3.19}$$

This element satisfies the continuity condition at the boundary for w but discontinuity occurs for rotations around the x and y axes (C^0-*continuity*). This element is thus referred to as the "*non-conforming*" element.

The 12 governing equations from each element can be written as,

$$\{\delta\}^e = [C]\{\alpha\} \tag{3.20}$$

where $[C]$ is a [12 x 12] matrix depending on the nodal coordinates and $\{\alpha\}$, a vector of 12 unknown constants. Inverting the equation,

$$\{\alpha\} = [C]^{-1}\{\delta\}^e \tag{3.21}$$

The expression for the displacement within the element can be expressed in terms of the shape function as,

$$\{f\} = w = [N]\{\delta\}^e = [P][C]^{-1}\{\delta\}^e \tag{3.22}$$

where,

$$[P] = \{1, x, y, x^2, xy, y^2, x^3, x^2y, xy^2, y^3, x^3y, xy^3\} \tag{3.23}$$

When the origin of the coordinate is at the center of the element, Eq.(3.22) can be expressed in terms of the shape function,

$$[N_i] = \frac{1}{2}[(\xi_0+1)(\eta_0+1)(2+\xi_0+\eta_0-\xi^2-\eta^2), \; a \cdot \xi_i(\xi_0+1)^2(\xi_0-1)(\eta_0+1),$$
$$b \cdot \eta_i(\xi_0+1)(\eta_0+1)^2(\eta_0+1)] \tag{3.24}$$

with,

$$\left.\begin{array}{ll} \xi = (x-x_i)/a, & \eta = (y-y_i)/b \\ \xi_0 = \xi \cdot \xi_i, & \eta_0 = \eta \cdot \eta_i \end{array}\right\} \tag{3.25}$$

The strains for the element can be given by,

$$\{\varepsilon\} = \left\{-\frac{\partial^2 w}{\partial x^2}, \; -\frac{\partial^2 w}{\partial y^2}, \; 2\frac{\partial^2 w}{\partial x \cdot \partial y}\right\}^T \tag{3.26}$$

$$= \left\{\begin{array}{c} -2\alpha_4 - 6\alpha_7 x - 2\alpha_8 y - 6\alpha_{11}xy \\ -2\alpha_6 - 2\alpha_9 x - 6\alpha_{10} y - 6\alpha_{12}xy \\ 2\alpha_5 + 4\alpha_8 x + 4\alpha_9 y + 6\alpha_{11}x^2 + 6\alpha_{12}y^2 \end{array}\right\} \tag{3.27}$$

$$= [B_b]\{\delta\}^e \tag{3.28}$$

Since,

$$\{\varepsilon\} = [Q]\{\alpha\} = [Q][C]^{-1}\{\delta\}^e$$

Thus $[B]$ can be expressed as,

$$[B_b] = [Q][C]^{-1} \tag{3.29}$$

in which,

$$[Q] = \begin{bmatrix} 0 & 0 & 0 & -2 & 0 & 0 & -6x & -2y & 0 & 0 & -6xy & 0 \\ 0 & 0 & 0 & 0 & 0 & -2 & 0 & 0 & -2x & -6y & 0 & -6xy \\ 0 & 0 & 0 & 0 & 2 & 0 & 0 & 4x & 4y & 0 & 6x^2 & 6y^2 \end{bmatrix} \tag{3.30}$$

Since certain terms in the diagonal axis of the matrix $[C]$ is zero, a direct inversion of the matrix is not possible. The $[Z]$ matrix method [3.10] is applied here to solve the problem.

The stiffness matrix for the plate bending element can be expressed as,

$$[k_b] = \int_v z^2 \cdot [B_b]^{T} \cdot [D] \cdot [B_b] \cdot dV \tag{3.31}$$

The full expression for the stiffness matrix can be given as,

$$[k_b] = \frac{1}{60ab}[L]\{D_x K_1 + D_y K_2 + D_1 K_3 + D_{xy} K_4\}[L] \tag{3.32}$$

The effects of coupling for moderately thick plates have to be considered especially for analytical studies of the ultimate behavior of concrete slabs[3.11 - 3.13]. At the ultimate stages, the amount of deformation can be large enough to cause failure for elements in the compressive regions. In this analysis, the effects of element coupling are considered by including the coupling related portions into the total element stiffness matrix,

$$[k_{pb}] = \int_v z \cdot [B_p]^{T} \cdot [D] \cdot [B_b] \cdot dV \tag{3.33}$$

$$[k_{bp}] = \int_v z \cdot [B_b]^{T} \cdot [D] \cdot [B_p] \cdot dV \tag{3.34}$$

The total element stiffness matrix can be then derived by adding up (super imposing) the terms in Eqs.(3.11), (3.30), (3.33) and (3.34) such as,

$$[k] = \begin{bmatrix} k_p & k_{pb} \\ k_{bp} & k_b \end{bmatrix} \tag{3.35}$$

The effects of transverse shear stresses in 2-dimensional plate bending problems are usually small enough to be ignored. But when dealing with analyses of multilayered composite structures and also thick plates, the effects of the transverse shear stresses have to be taken into consideration as delamination and failure through the slab are enhanced by the stresses[3.7, 3.11 - 3.14]. It is a common practice to classify thin plates from thick plates by considering the ratio of slab thickness (h) to the characteristic length (a). From results of nondimensional analysis, it is found that under normal circumstances, the inplane stresses σ_x, σ_y and τ_{xy} are of the order of "qa^2/h^2", whereas the transverse shear stresses τ_{xz} and τ_{yz} are of the order of "qa/h", and the transverse stress σ_z is of the order of "q", where "q" is the external load [3.12].

The dimensions of the concrete slabs considered here are 130.0 x 130.0 x 13.0 cm. Slabs with a thickness to characteristic length (h/a) of less than 1/10 are usually considered as thin plates while those with a larger ratio are usually referred to as thick plates. The slabs considered here have a "h/a" ratio of exactly 1/10 and thus the effects of the transverse shear stresses may not be small enough to be ignored. Furthermore, the dimension of concrete handrails considered is 400.0 x 107.5 x 25.0 cm. It is considered that the effects of these transverse shear stresses become more apparent during impact or impulsive loadings and is one of the dominant factors that contribute towards punching shear failure under high loading rates.

A sure and certain method of deriving the transverse shear stresses is to apply a 3-dimensional constitutive equation and solving for the transverse shear stresses and strains. But it is considered here that a complicated constitutive equation might not be very applicable for cases of impact or impulsive loading and application of a complicated constitutive equation holds not much merit. A simple and relatively inexpensive method is used here, where the stresses are calculated from the equilibrium equations of a 3-dimensional elastic body with the transverse stress $\sigma_z=0$. The transverse shear stresses τ_{yz}, τ_{xz} for 2-dimensional plates can be taken into account by considering the following equilibrium equations,

$$\frac{\partial \sigma_x}{\partial x} + \frac{\partial \tau_{xy}}{\partial y} + \frac{\partial \tau_{xz}}{\partial z} = 0 \tag{3.36}$$

$$\frac{\partial \sigma_y}{\partial y} + \frac{\partial \tau_{xy}}{\partial x} + \frac{\partial \tau_{yz}}{\partial z} = 0 \tag{3.37}$$

The transverse shear stresses can be obtained by direct integration of the equation of equilibrium,

$$\tau_{xz} = -\int \left(\frac{\partial \sigma_x}{\partial x} + \frac{\partial \tau_{xy}}{\partial y} \right) \cdot dz \tag{3.38}$$

$$\tau_{yz} = -\int \left(\frac{\partial \sigma_y}{\partial y} + \frac{\partial \tau_{xy}}{\partial x} \right) \cdot dz \tag{3.39}$$

Using the method mentioned above, the stress-free boundary condition for the transverse shear stresses are not satisfied. The following boundary conditions are included here to satisfy the stress-free condition at the upper and lower faces,

$$\tau_{xz}(x, y, \pm h/2) = \tau_{yz}(x, y, \pm h/2) = 0 \tag{3.40}$$

where, h is the thickness of the slab.

The stress-free boundary condition can be included into the equation by introducing a function $f(x, y)$ such as,

$$\tau_{xz} = -[z] \cdot \left(\frac{\partial \sigma_x}{\partial x} + \frac{\partial \tau_{xy}}{\partial y} \right) + f_1(x,y) \tag{3.41}$$

$$\tau_{yz} = -[z] \cdot \left(\frac{\partial \sigma_y}{\partial y} + \frac{\partial \tau_{xy}}{\partial x} \right) + f_2(x,y) \tag{3.42}$$

A determinate solution for the above equation can be found by replacing the stresses by moments,

$$\sigma_x = \frac{12M_x \cdot z}{h^3}, \quad \sigma_y = \frac{12M_y \cdot z}{h^3}, \quad \tau_{xy} = \frac{12M_{xy} \cdot z}{h^3},$$

(3.43)

Substituting into the equilibrium equations above,

$$\tau_{xz} = -\frac{6z^2}{h^3} \cdot \left(\frac{\partial M_x}{\partial x} + \frac{\partial M_{xy}}{\partial y}\right) + f_1(x,y)$$

(3.44)

$$\tau_{yz} = -\frac{6z^2}{h^3} \cdot \left(\frac{\partial M_y}{\partial y} + \frac{\partial M_{xy}}{\partial x}\right) + f_2(x,y)$$

(3.45)

Introducing the boundary conditions $\tau_{xz} = \tau_{yz} = 0$ at $z = \pm h/2$ into the above condition and solving for $f_1(x, y)$ and $f_2(x, y)$,

$$\tau_{xz} = -\frac{3}{2h} \cdot \left(\frac{\partial M_x}{\partial x} + \frac{\partial M_{xy}}{\partial y}\right)\left(1 - 4 \cdot (\tfrac{z}{h})^2\right)$$

(3.46)

$$\tau_{yz} = -\frac{3}{2h} \cdot \left(\frac{\partial M_y}{\partial y} + \frac{\partial M_{xy}}{\partial x}\right)\left(1 - 4 \cdot (\tfrac{z}{h})^2\right)$$

(3.47)

The partial derivatives ($\partial \sigma_x/\partial x$, $\partial \sigma_y/\partial y$, $\partial \tau_{xy}/\partial x$, $\partial \tau_{xy}/\partial y$) in the above equations have to be determined in order to calculate the transverse shear stresses. Here, the stresses are calculated at the integration point (point P_0 in Fig.3.6) as well as at very nearby points on either side of the integration point in the direction of the derivative variable (points P_1, P_2, P_3, P_4 in Fig.3.6). These three points are then used to determine the three constants of a second-order polynomial that can be used to obtain the required derivative (Illustrated in Fig.3.7).

An example of the stress distributions in a reinforced concrete slab is shown in Fig.3.8. It can be noticed that the transverse shear stresses are dominant towards the center (middle) of the slabs.

3.1.2 LAYERED APPROACH FOR SLAB STRUCTURES [3.1, 3.3, 3.4, 3.12, 3.13]

Reinforced concrete slab structures are modeled using the layered approach. Fig.3.9 shows a diagrammatical illustration of the layering concept employed here. The slab structure is divided into a few discrete layers in the thickness direction, both of concrete and of reinforcement. The layering approach allows the strains and stresses to be varied with member thickness and permits the inclusion of the steel reinforcement at the proper level within the slab.

From Fig.3.9, it may be seen that the simulated composite reinforced concrete element resembles a laminated plate element. This method allows different material properties to be assumed for each layer. Due to the nonhomogeneous nature of the composite structure and also the possibility of unsymmetrical locations of the different laminates with respect to the midplane of the element (reference surface - Refer Fig.3.10), the in-plane extension and transverse bending of the elements are usually coupled.

During nonlinear analysis of reinforced concrete slabs, stress-induced anisotropy in concrete layers due to the progression of cracks through the layers cause the neutral axis to shift across the cross-section. In such cases, the bending-extensional coupling becomes even more pronounced due to the shifting of the neutral axis.

The main assumptions for the layered approach can be summed up as follows:

(1) The slab is considered to be made up of hypothetical reinforcement layers, which resist axial and in-plane shear forces, and concrete layers. The in-plane shear forces in the reinforcement can be considered by including the values for related terms in the material stiffness matrix.

(2) Strain in the reinforcement and concrete layers are assumed to be proportional to the distance from the neutral axis (Refer Fig.3.11). Concrete layers are in the state of plane stress and there is no slip between layers.
(3) Concrete is considered to be orthotropic after cracking. The amount of strain energy in the element is converted into equivalent nodal forces after cracking. The numerical representation of cracking is based on a 2-dimensional smeared crack approach in the in-plane direction, where the effects of aggregate interlock and dowel action after cracking can be expressed in terms of a shear retention factor.
(4) Material properties obtained from static uniaxial tests are converted into an effective stress-effective strain relation in the multiaxial failure model for concrete and used as the input data. Uniaxial material characteristics are applied for reinforcement.
(5) Failure is defined in the analysis as the point where either concrete crushing under compression or failure of reinforcement occurs in a structural element. Three classifications of failure modes are defined in the analysis based on the deformation mode, impact force-deflection relation, failure conditions in the elements and also crack patterns. The failure modes are; (i) Bending failure, (ii) Bending to punching shear failure, and (iii) Punching shear failure. In the bending to punching shear failure mode, the bending mode is dominant in the earlier stages of loading and is then followed by a transition to the punching shear mode (Refer Sect. 3.3.2).

The total strains for each layer are considered to consist of the membrane element strains $\{\varepsilon_0\}$ (Eq.(3.9)) and the plate bending related strains $\{\varepsilon\}$ (Eq.(3.28)). The membrane strains are constant throughout the thickness of the slab but the strains caused by plate bending varies as the layer separates from the middle surface (reference surface). The total strains for each layer can thus be expressed as,

$$\{\varepsilon_T\} = \{\varepsilon_0\} + z \cdot \{\varepsilon\} \tag{3.48}$$

where z is the distance from the reference surface.

The neutral axis is assumed to act at the center of the elastic portion of the slab, expressed by the following equations,

$$e_x = \frac{(1/2)E_c \cdot t^2 + E_s \cdot \Sigma A_{sxi} \cdot Z_{sxi}}{E_c \cdot t + E_s \cdot \Sigma A_{sxi}} \tag{3.49a}$$

$$e_y = \frac{(1/2)E_c \cdot t^2 + E_s \cdot \Sigma A_{syi} \cdot Z_{syi}}{E_c \cdot t + E_s \cdot \Sigma A_{syi}} \tag{3.49b}$$

where,
   e_x, e_y: the neutral axes in the x and y directions, respectively.
   E_c, E_s : concrete and reinforcement moduli of elasticity, respectively.
   A_{sxi}, A_{syi}: average cross-section per unit length in the x, y directions for the i th layer of reinforcement, respectively.
   Z_{sxi}, Z_{syi}: distance from the middle of i th layer to the top surface of slab in the x, y directions, respectively.
   t: slab thickness.

The total stiffness matrix for the composite element is obtained by integrating the stiffness matrix for each layer and summing up for the total stiffness,

$$[k]_T = \sum_{i=1}^{l} [k]_i \tag{3.50}$$

where l is the total number of layers in the layered approach.

The question of element failure is a common problem when dealing with brittle materials such as concrete. The finite element method is basically applicable only for a continuum. But when dealing with finite element analysis of concrete materials, the discontinuity in the continuum caused by cracks has to be taken into consideration.

Element failure (crushing and cracking) in the concrete element is overcome by dissipating the relevant elastic strain energy to the surrounding element nodes in the form of equivalent nodal forces. The equivalent nodal forces can be expressed as,

$$\{\overline{F_t}\} = \int_v [B]^{T\cdot} [T_{\varepsilon\theta}]^{T\cdot} \{\sigma_\theta\}^\cdot dV$$

(3.51)

where,

$\{ \overline{F_t} \}$: equivalent nodal forces.

$[B]$: element $[B]$ (internal stress-nodal displacement) matrix.

$[T_{\varepsilon\theta}]$: transformation matrix relating global directions to crack direction.

$\{\sigma_\theta\}$: stresses in the θ direction where the stress in the direction perpendicular to θ is set to zero.

θ : direction of main principal stress.

Fig.3.12 shows the model and orientation of cracking in the concrete element.

The associated terms in the material stiffness matrix are reduced to zero for elements with cracks. For plane stress problems, the material stiffness matrix after cracking in the main principal stress direction can be given by,

$$[D_{ep}] = \frac{E}{1-v^2} \begin{bmatrix} 0 & 0 & 0 \\ 0 & 1 & 0 \\ 0 & 0 & \alpha^\cdot (1-v)/2 \end{bmatrix}$$

(3.52)

Another problem associated with concrete cracking is the transfer of shear forces at the crack surface mainly caused by *aggregate interlock* (AI). Normally, the amount of contribution due to shear transfer can be expressed in terms of the crack width and also crack spacing, and there are some models that are capable of dealing with this phenomenon [3.15, 3.16]. The amount of shear transfer decreases as the crack width increases, due to the separation in the surfaces. But for convenience, a constant value is adopted here. The effects of shear transfer at crack surfaces is included in the matrix in the form of α, which is often referred to as the *shear retention factor*. The shear retention factor is set at 0.3 for the calculations in this study.

3.1.3 MODELING OF CONCRETE SLAB STRUCTURES

Two different types of concrete slab structures are modeled using the layered finite element procedure. Reinforced concrete slabs (130.0 x 130.0 x 13.0cm) and reinforced concrete guardrails (handrails) (400.0 x 107.5 x 25.0cm) with doubly reinforced section are modeled as shown in Figs.3.13 and 3.14, respectively. The reinforced concrete slabs are simply supported on two edges while the reinforced concrete handrails have fixed supports on one side. The slabs represent a normal slab structure while the handrails are usually used on expressways and can also represent a concrete barrier wall.

The external load for the reinforced concrete slab in Fig.3.13 is applied as a distributed patch (12.0 x 12.0cm) load at the center of the slab while a similar distributed patch (60.0 x 10.0cm) load is applied at a height of about 72.5cm from the bottom fixed supports for the concrete handrail, as indicated in Fig.3.14. The loading position in the concrete handrail is meant to simulate the effects of collision from a vehicle. Due to the symmetrical layout and also external conditions, only a 1/4 portion of the slab and a 1/2 portion of the handrail are discretized in the finite element idealization.

3.2 VERIFICATION OF ANALYTICAL PROCEDURE

3.2.1 EXPERIMENTAL PROCEDURE

The validity of the analytical method is verified by means of comparison with available experimental data. Full scale reinforced concrete slabs with a dimension of 130.0 x 130.0 x 13.0cm are subjected to both static and impact failure tests[3.17]. The static tests are carried out in order to be able to distinguish the difference in failure modes and failure conditions as a comparison to impact loading. The dimensions and details of the slabs are as shown in Fig.3.15.

For the static tests, load is applied through a 50tf capacity hydraulic jack, and is gradually increased by an increment of 1.0tf. At each loading step, the load is held constant to measure automatically the deflection, concrete and reinforcement strain and crack width. The displacement at the center of the slabs are recorded by an X-Y recorder.

On the other hand, the apparatus used for the impact loading test is a pendulum type impact testing machine which is specially designed to derive only one sine-wave impact (Fig.3.16). The falling weight has a mass of 500kgf. In order to derive soft impacts, a rubber pad (1.0cm thick) is placed on a square steel loading plate (15.0 x 15.0 x 1.5cm) at the impact face. The impact load-time relations are measured by acceleration sensors attached to the falling mass. Measurements for deflection, acceleration response and crack widths are carried out. The measuring system consists of non-contact displacement transducers, acceleration sensors, crack gauges and an analog data recorder. The measured points are shown in Fig.3.17.

The height of fall of the impacting mass for the failure tests are first estimated by means of the analytical procedure. Based on the assumption that the failure energy is totally transferred to the slab during impact, the height of fall is estimated. Fig.3.18 shows the definition of impact failure energy in the slabs. The impact failure energy for condition "2" is selected for the tests and the height of drop of the mass is predicted. Furthermore, tests in the elastic region are also carried out, before the ultimate failure tests, by dropping the mass from a height of 1.5cm.

The impact load function measured during experiments are digitalized using an *analog-to-digital* (A-D) transformation process by means of a mini computer and is then input into the analysis. Material test results such as *Young's* modulus, *Poisson's* ratio, uniaxial material characteristics from uniaxial compressive (concrete) and tensile (concrete and reinforcement) tests are used as input data for the materials. It should be noted here that the uniaxial material test results are converted into the effective stress-effective strain relation before application into the calculations.

The test program consists of an impact loading test on full-scale reinforced concrete handrails [3.18, 3.19]. The details of the test setup are as shown in Fig.3.19 while a cross-section diagram of the concrete handrail is illustrated in Fig.3.20. Three cranes are used to induce the impact force as shown in the figure. Two cranes are used to suspend the 2.5tf impacting mass, one on each side, in order to fix the collision course (movement) of the impacting mass, while another crane is used to raise the mass to the proper height of fall. The mass is placed in such a way that it would act similar to a striking pendulum, acting directly perpendicular to the face of the handrails. A metal loading plate is placed at the face of the handrails to induce the impact load equally through a rectangular surface (distributed patch load). The height of fall for the mass is set at 4.0cm (Potential energy = 0.10$tf•m$) for the elastic tests and 105.2cm (Potential energy = 2.63 $tf•m$) for the failure tests. After the initial failure tests, the height of fall is set at 90.0cm (Potential energy = 2.25$tf•m$) for further tests to check the progress of deterioration after initial failure. The measuring system consists of eddy current type non-contact displacement transducers, the *Opto-Follow*, which is an optical instrument for measuring deflections remotely, acceleration sensors and an analog data recorder. The measuring setup is shown in Fig.3.21.

3.2.2 EXPERIMENTAL RESULTS AND DISCUSSIONS

Results of experimental tests carried out on reinforced concrete slabs and handrails are used to verify the validity of the calculations. Results of tests carried out on full-size concrete slabs and handrails of the same dimensions as those in the analysis will be used here. The types of concrete slabs tested are the normal strength reinforced concrete (RC) slabs, high strength reinforced concrete (HRC) slabs, steel fiber reinforced concrete (SFRC) slabs, double layered reinforced concrete slabs and fiber reinforced plastic (FRP) reinforced concrete slabs.

Firstly, the results for experiments and analysis for impact loading in the elastic region are considered. Fig.3.22 shows the impact force-midspan deflection relation for experiment and analysis of a normal reinforced concrete (RC) slab. The analysis predicts the impact force-midspan deflection very accurately, especially the vibrational effects in the curves. Fig.3.23 shows the corresponding comparison between experiment and analysis of the midspan acceleration response in relation to time. The analysis predicts the response relatively accurately with a frequency similar to the experimental results.

The results herewith are performed using the *Ottosen* triaxial failure model together with a triaxial yield criterion. Furthermore, the effects of unloading and also transverse shear stresses are incorporated into the analysis.

Fig.3.24 shows the impact load versus midspan deflection curves for the calculations and experiments of two different RC slabs, i.e., slab (RC-D1) and slab (RC-D3). The height of fall for the $500kgf$ impacting mass is indicated by the notation "h". In the first test, the height of fall is set at $h=30.0cm$ while the height of fall for another different concrete slab is set at $h=60.0cm$. The results of Fig.3.24 show that the calculations give a very good approximation of the ultimate behaviors of RC slabs especially until the point of maximum impact load. The difference between the calculated values and the experiments begin to appear after the maximum impact load, i.e., when the unloading process begins. A slight difference can be noticed in Fig.3.24(a) in the initial elastic stages, but it can be concluded here that the analysis gives a good prediction of the real behavior as it is a usual phenomenon for the curve to be parabolic in the initial stages during impact loadings, due to inertial effects. The larger deflection in the tested slabs can be attributed to small gaps at the supports of the tested slabs. Fig.3.24(b) shows a better prediction on the overall, even in the unloading stages except for slight vibrational effects towards the end of the analysis.

The deformation mode at failure for both concrete slabs obtained from the analysis are shown in Fig.3.25. An overall deformation can be noticed in slab (RC-D1), where total structural failure is expected. The failure mode in this case is the bending failure mode. When the height of fall is increased, as in slab (RC-D3), the loading rate for the impact load function also increases. This causes local failure to be more evident, as noticeable in the middle of slab (RC-D3) in Fig.3.25(b). The failure mode in this case is considered to be the bending to punching shear failure mode, where the bending deformation is dominant in the earlier stages (not indicated in the figure) and then followed by a transition into the punching shear mode at the final stages.

Fig.3.26 shows the comparison of crack pattern at failure for the analysis and test of slab (RC-D3). The analytical results show the direction perpendicular to the main principal stress at the bottom (rear) layer (8th layer) of the slab, giving an indication of the cracking pattern. In the analysis, the cracks basically radiate from the center of the slab towards the edges. Cracks perpendicular to these cracks also appear in the form of a circle, giving an implication of punching shear failure, which is the failure mode noticed in this experiment.

The impact force versus midspan deflection curves for the calculations and experiments of two different high strength reinforced concrete (HRC) slabs are shown in Fig.3.27. In the first test, the height of fall is set at $h=60.0cm$ while the height of fall for another different concrete slab is set at $h=80.0cm$. The slabs are denoted as HRC-D2 and HRC-D3, respectively. Based on the results shown in Fig.3.27, it is clear that the calculated results are a little different from the experiments after the initial cracking stages. But on the whole, the calculations are able to give a rough estimate of the overall behavior. In Fig.3.27(a), the calculations give a larger value of deflection while a smaller value of deflection is predicted in Fig.3.27(b). The reason for the difference here can be assumed to the fact that the *Ottosen* failure model is not capable of predicting accurately the ultimate behaviors of high strength concrete. The *Ottosen* model was initially developed and calibrated only for normal strength concrete and as such is not really applicable here. But on the whole, it still gives a good estimate of the overall response under impact loadings. It is considered here that a few alterations are still necessary in the failure model for it to be able to be applied for analysis on high strength reinforced concrete, especially to the point where initial cracking is expected to occur at a higher strain value compared to normal strength concrete.

New innovations in the field of civil engineering are usually related to innovations of materials or in a structural sense, such as a more durable structure, lightweight structure, aesthetic structure, etc. The development of fiber reinforced plastic (FRP) based materials as a substitute for concrete reinforcement is one such example. It does not cover only the material aspect of the problem, but further innovations in the structural sense can also be expected due to the decrease of cover concrete, and thus indirectly decreasing the dead weight too. Furthermore, due to the anti-corrosive nature of the FRP materials, durability is one major point where positive steps forward could be expected. The FRP based reinforcements are also known to be resistant to chemical corrosion. Since it also bears no influence on magnetic fields due to its non-magnetic conductivity, it is considered ideal for use on structures carrying high technology magnetic elevated trains in the near future. But there are also a few set backs, such as the problem of bond between the FRP reinforcement and concrete, a relatively low modulus of elasticity and also because there is no yield or plastic phase in the FRP materials. Therefore, there is still a necessity of providing an adequate amount of safety factor for structures using FRP materials[3.21 - 3.23].

Before a practical application of FRP materials to structures can be allowed, studies on the behavior of structures under static and dynamic loads, as well as evaluations of the related performances have to be established. Moreover, there has been a growing trend of using concrete for construction in various fields. With a wider application of concrete for structural use, especially to structures of public importance, the problem of impacts caused by natural collisions or accidents onto concrete structures have to be given proper considerations too. In order to check the applicability and validity of the layered finite element procedure for analysis of FRP reinforced concrete slabs, experiments on full scale test specimen are performed.

The types of fiber composite materials usually used for reinforcing concrete can be grouped as glass fibers (G), carbon fibers (C) and aramid fibers (A). The strands of the fibers are twisted and embedded in resin materials. Two different carbon based FRP reinforcements (CFRP reinforcements) are used in the experiments. Table 3.1 shows the mechanical properties of the two types of FRP reinforcement together with the normal steel reinforcement bar. Since the FRP reinforcements have a relatively low modulus of elasticity, cracking in concrete is considered to occur at an early stage. In order to supplement this demerit, a combination with high strength concrete (HRC) is also considered. The material characteristics of the concrete types are shown in Table 3.2. Five different slabs are tested, i.e., normal strength concrete with steel reinforcement (RC-35), normal strength concrete with FRP type 1 (RC-1), high strength concrete with FRP type 1 (HRC-1), normal strength concrete with FRP type 2 (RC-2), and high strength concrete with FRP type 2 (HRC-2). The main results from tests on the slabs are shown in Table 3.3.

The comparison of impact force-deflection curves for the experiment and analysis of the RC-35 slab is shown in Fig.3.28. Fig.3.28(a), (b), (c) and (d) show the deflection functions at midspan of the slab, quarter span of slab in the transverse direction, quarter span of slab in the longitudinal direction and edge of the slab in the longitudinal directions, respectively. This is in relation to points 1, 2, 3, 4, respectively in Fig.3.17. On the whole, the analysis gives very good predictions of the deformation behaviors at all points, even in the unloading regions. But some vibrational effects in the analysis are noticed at the end of the calculations due to instability in the analysis as failure progresses rapidly. Therefore, the failure progress can be considered to be slightly faster in the analysis. The analysis predicts the deformation behavior at the edge of the slab (Fig.3.28(d)) very accurately, especially the rapid increase in deformation after reaching an impact force of about 25tf.

Similar impact force-deflection curves at the same four points for the experiments and analysis of the RC-1 and HRC-2 slabs are shown in Figs.3.29 and 3.30, respectively. Generally, the analysis gives a rough prediction of the impact force-deformation behaviors at all four points. But comparatively, the accuracy is not as high as in the RC-35 slab. The main reason is due to the difference in bond behavior between reinforcements and concrete. In the RC-35 slab, the bond between steel reinforcement and concrete is relatively higher because the reinforcement bar is deformed and there is a large yielding plateau for steel. In the layered finite element analysis applied, it is assumed that the bond between reinforcement and concrete is perfect. But in the FRP reinforcements, the surface of the reinforcement is smooth and not much bond can be expected. Furthermore, the FRP reinforcements are fabricated in a 2-dimensional grid, with the central axis of reinforcement in both the transverse and longitudinal directions on the same point. Therefore, bond failure and also bond cracks propagates at an early stage, mainly in the transverse directions, but the stresses are immediately transferred to the longitudinal directions at the intersections of the reinforcement bars. Therefore the crack patterns are totally different in both cases. In the FRP reinforced concrete slabs, the cracks mainly run along the reinforcing bars, in both the transverse and longitudinal directions but on the other hand, the cracks radiate out from the center of the slab for the RC-35 slabs.

In the analysis carried out, the failure progress is faster and also highly localized in the FRP reinforced slabs due to the bond effects. But in reality, the 2-dimensional bond failure spreads the region of failure throughout the slabs andcauses total damage to be global. But on the whole, the analysis still gives relatively good predictions of the impact force-deflection curves. Furthermore, the improvement in failure modes from punching shear to bending by application of HRC is correctly predicted in the analysis. Therefore, it is considered that the layered finite element procedure could be practically applied for analysis of FRP reinforced concrete slabs. But a model that could include the bond effects in the reinforcement would further increase the accuracy of the analysis.

Verification of the analysis for reinforced concrete guardrails (handrails) are also performed by comparisons of deformation and acceleration responses[3.18, 3.19, 3.24]. Fig.3.31 shows the comparison of the experiment and calculation for deflection at the loaded point in relation to time. The calculations predict a slightly smaller deflection but the curve shapes are accurately predicted in the analysis. The comparison between experiments and calculations for the acceleration response with respect to time at the loaded point is shown in Fig.3.32. On the overall, the curve shapes as well as the acceleration values are accurately predicted in the analysis. The frequency is also accurately predicted in the analysis. Fig.3.33 shows a comparison of the expected crack patterns in the calculation and experiment. Fig. 3.33(a) and (b) shows the direction perpendicular to the main principal stresses in the first (front) and last (rear) layers at two different times, respectively. The final crack pattern shown in Fig.3.33(c) is accurately simulated as shown in Fig.3.33(b).

3.3 ANALYSIS OF FAILURE MODES [3.25 - 3.27]

Present day design codes in most countries are based mainly on static design methods. The effects of dynamic loads are usually taken into consideration by introducing an equivalent static load factor. The same case also applies to the design of impact resistible structures where usually an impact factor is introduced into the static design method. The dynamic forces presumed to occur during an impact collision is then converted into a static force of equal magnitude and treated in much the same way as other static loads. This static design method would not adequately describe an impact phenomenon and thus have only a limited amount of practical applicability, especially in the case of concrete structures. During an impact collision, an excitation of not only the first mode but also other higher modes can be expected. Structures designed using the equivalent static force would be able to withstand bending, but not shear, hence bringing about punching shear or concrete scabbing in the rear face. These factors can only be totally considered if a dynamic design approach is adopted.

The analysis of the failure modes of reinforced concrete slabs under soft impact loading will be carried out here. The main intention of this Section is to identify the various failure modes during soft impact loadings with the further intention of applications to design of concrete slab structures. In the design of concrete structures, there is a necessity to prevent shear failure from occurring as shear failure is considered to be more hazardous. Structures are usually designed to fail by bending in the various design codes. Moreover, a bending failure mode would reduce the possibility of scabbing at the rear face to occur.

3.3.1 FAILURE MODES FOR CONCRETE SLAB STRUCTURES UNDER IMPACT LOADS

The process where a concrete structure is subjected to an impact load is shown in Fig.3.34 [3.28]. The impact causes pressure, which is a function of time, which in turn causes stresses to occur. The stresses are the main cause of failure. It is considered here that the type of failure mode occurring depends on the pressure or load-time function generated during the impact collision.

Impacts caused by collisions can normally be classified into two different types, i.e., hard impacts and soft impacts. The failure modes generally occurring during soft impacts are either bending failure or punching shear failure. This study is limited to the region of soft impacts and thus the main purpose of this Section is to distinguish analytically the difference between bending failure and punching shear failure.

3.3.2 ANALYTICAL DEFINITION OF FAILURE MODES

The failure modes under soft impact loadings will be analytically defined here based on the various analytical results and also keeping in mind the results from experiments carried out. It is considered here that the failure modes can be defined based on the following indexes;

(1) The distribution of deflection along the transverse and longitudinal directions (deformation mode),
(2) The impact load - midspan deflection curves from zero to ultimate load,
(3) The progression of cracks in the transverse and longitudinal directions,
(4) The crack pattern at failure.

Fig.3.35 shows the results by F. Gonzallez-Vidosa et al. [3.29] where the failure modes for static loading of symmetrical reinforced concrete slabs are investigated using the 3-dimensional finite element method. The analytical prediction of failure modes in this study is based on;
(1) The maximum load-carrying capacity of the slab (ultimate load),
(2) The crack pattern,
(3) The load-deflection curves from zero to ultimate load.

From Fig.3.35, it is clear that the failure modes for static loading of symmetrical reinforced concrete slabs can be predicted from the load-deflection curves.

It is considered here that the failure modes of concrete slabs are related to the loading rates of the impact load-time functions. For a given amount of impulse, the failure modes would change with the loading rates. Therefore, the results of the four indexes mentioned above will be considered together with the effects of the loading rate.

In order to be able to predict the failure modes of concrete slabs, there is a necessity to define failure in the calculations carried out. It is considered here that the failure modes are phenomena involving the total behavior of the slabs (structural behavior) while failure is taken as the local behavior of the slabs (material behavior). In the analysis carried out, failure is considered to occur either when one of the following conditions are satisfied;
(1) Concrete fracture occurs under compression (concrete crushing),
(2) Tensile rupture of reinforcement,
(3) The point where deflection decreases even with the increase of impact load.

(1) Distribution of deflection
The distribution of deflection in the transverse and longitudinal directions are shown in Figs.3.36. The loading rate increases from (a) to (c). In the figures, a parabolic curve with a gradual and smooth surface (Fig.3.36(a)) is considered to be that of bending while curves with a high order function or with a sudden change in gradient (with a point of inflection) (Fig.3.36(c)) is considered to be that of shear failure. The solid lines show the deflection at failure while the broken lines are those at a stage before failure.

Fig.3.36 (a) can be grouped as those of typical bending failure. Deflection is spread at an equal ratio throughout the cross-sections. It can be considered that the energy from the impact load is spread throughout the entire structure and only total structural failure is expected. In Fig.3.36(b), the bending failure mode is noticed at a stage before failure followed by a quick transition into the punching shear failure mode. On the whole, it is considered that punching shear failure occurs but if the amount of impulse of the loading function is reduced, then the progress of failure would stop at the earlier stage shown and the failure mode would then be bending failure. Fig.3.36(c) shows the results of very fast loading rate. The punching shear failure mode appears from a very early stage and it can be noticed that there is almost a negligible amount of deflection at the sides of the slabs. Therefore, in this case, local failure occur where only the middle portion of the slabs are affected. It can be concluded that the distribution of deflection gives a good idea of the failure modes in concrete slabs.

(2) Impact load - midspan deflection curves
Fig.3.37 shows the analytically obtained impact load versus midspan deflection curves for three different ranges of loading rates. The range of the loading rate increases from (a) to (c). The scale for the abscissa of the figures are kept constant so as to be able to compare the different failure modes. The curves for bending failure can be assumed to have a smaller gradient than those of punching shear followed by a rapid increase in deflection at the ultimate stages due to the yielding of reinforcement throughout the slab (Fig.3.37(a)). As for punching shear failure, the initial gradient is smaller and not much deflection can be expected at failure (Fig.3.37(c)). Fig.3.37(b) can be classified as that of the bending to punching shear failure type.

Fig.3.38 shows the combined results of all the calculations in a single figure. The impact load at failure increases with the loading rate. Furthermore, an increase in initial stiffness can be noticed as the loading rate is increased, and these results are in good agreement with general impact phenomena. These effects can be attributed mainly to inertia of the structure. From the results, a failure envelope in the shape of a parabolic curve is obtained as shown in the figure by dotted line. The impact load at failure increases with increasing rate of loading, but the corresponding deflections at both high loading rates and low loading rates are greater than at the medium loading rates. This fact is of importance as it indicates that the *failure mechanism and also failure modes are different for different regions of loading rate*.

(3) Progression of cracks

Fig.3.39 shows the progression of cracks at about 80% of failure load in the cross-section of slabs in the transverse and longitudinal directions. The results for two extreme cases are shown. For the lower loading rate (Fig.3.39(a)), the area affected by cracks spread over a wide area, with radial cracks forming mainly at the sides of the slab. In Fig.3.39(b), the cracks are concentrated only at midspan and the cracks progress from the bottom and top surfaces. Not much radial cracks are noticed in this case. In other words, for punching shear failure as in Fig.3.39(b), local failure at midspan becomes more pronounced while during bending failure (Fig.3.39(a)), overall structural failure can be expected.

(4) Crack pattern at failure

Fig.3.40 shows the crack patterns at failure for slabs of two different loading rates. The crack pattern mentioned here is the direction perpendicular to the principal axes of the elements at the rear face (8th. layer) of the slab. The slabs show not much difference in the crack patterns. Basically, most cracks radiate from the middle of the slab towards the outer edges. Cracks perpendicular to these cracks also appear near the edges of the slabs forming in the shape of a circle. The only noticeable difference in the crack patterns is that the circular cracks seem to grow smaller as the loading rates increase. This would mean that the area affected by failure becomes smaller at higher loading rates and also an implication of the region to be affected by punching shear failure.

From the above results, all the 4 indexes show a good estimate of the failure mechanism and also the failure modes. Index (1) and (2) gives the most reliable prediction of the failure modes on the whole. A summary of the types of failure modes and the related physical behaviors are shown in Figs.3.41 and 3.42 [3.22].

(It should be mentioned that the loading rate of experiments carried out presently are limited to about 20*tf/msec* due to the capacity of the testing apparatus. The results within this range seem to agree with the calculations but the results of analysis of fast loading rates have yet to be experimentally verified. But the results from the analysis show a good agreement with what is generally expected under impact loadings.)

3.4 DYNAMIC RESPONSE OF CONCRETE SLAB STRUCTURES

The ultimate dynamic behaviors and also failure modes in reinforced concrete slabs under different conditions are studied analytically. Firstly, the effects of transverse shear stresses and also the unloading phenomena during impact loading are considered. Next, a study on the effects of different loading rates on different behaviors is performed. Finally, material considerations such as concrete type and effects of FRP reinforcement are considered.

3.4.1 EFFECTS OF TRANSVERSE SHEAR STRESSES

Results mentioned here are based only on analytical results as there is no experimental data available on the effects of transverse shear stresses. Therefore, there is a need to carry out further verifications in the future by carrying out the necessary measurements during experiments. The transverse shear stresses are zero at the free surfaces and attain a maximum value at the middle of the slabs in a parabolic shape. Therefore, the transverse shear stresses mainly cause internal cracks to occur in the region towards the middle of the slabs. It is difficult to experimentally check the real phenomenon during experiments as loading under impacts are usually of the transient type and the duration of loading is very small.

Fig.3.43 shows an example of calculations with and without the effects of transverse shear stresses compared to the experimental results. In this case, not much difference can be noticed from both calculations except in the later stages of the calculations. The calculations considering the effects of the transverse shear stresses seem to give a more accurate prediction of the curve in the unloading stages.

3.4.2 EFFECTS OF UNLOADING [3.31]

The effects of unloading for analysis of structures under impact loading are necessary as the impact load function is very complex in itself and the effects of inertia during high loading rates can cause failure to occur even when the whole structure is in the unloading stages. For design of structures under impact loadings, it is necessary that the structure withstand the whole impact load function. The impulse from the whole loading function plays a big role in the behavior of the structure and consideration of the maximum impact load only, as in the static case, is not adequate.

Besides that, another phenomenon in the dynamic analysis of concrete structures is the possibility of local unloading in certain elements even though the whole structure is in the loading process. This is mainly caused when the elastic strain energy is released at the nodal points due to concrete fracture or cracking. The equivalent nodal forces released at the nodes can cause local unloading in the adjacent elements. Therefore, it can be concluded that a dynamic analysis for brittle materials without considerations for the unloading process is improper, even though unloading of the whole structure is not considered.

Fig.3.44(a) shows an example of the impact load versus time function obtained from experiments on RC slabs. The corresponding calculated results for the midspan total concrete strain at the bottom surface of the RC slab is shown in Fig.3.44(b). The strains increase linearly in the initial stage until the formation of cracks, where the strains increase at a faster pace. It is clear here that even when the whole structure is under the unloading process, the strains still increase at a very fast pace. This can be attributed to the effects of inertia during impact loadings. The strains only begin to decrease gradually after the impact load has decreased to about 23 *tf*. Therefore there is a possibility of failure progressing rapidly during the unloading process for impact loadings. The conclusion that can be obtained here is that for analysis on impacts of structures, a dynamic analysis until the maximum impact load only is not adequate for the study of the ultimate behaviors and prediction of failure modes.

3.4.3 EFFECTS OF LOADING RATE [3.26, 3.27, 3.32]

Analytical results of the various ultimate behaviors of RC slabs and its relation to the loading rates will be carried out here. The main results from the analysis are shown in Table 3.4 while the diagrammatic illustration of the results are shown in Figs.3.45 to 3.49. The factors considered here are:
   (1) Cracking load,
   (2) Reinforcement yielding load,
   (3) Concrete plasticity (compression) load,
   (4) Impact force at failure,
   (5) Deflection at failure,
   (6) Impulse from the impact load function,
   (7) Energy absorbed.

Fig.3.45 shows the effects of loading rate on cracking, reinforcement yielding and plastic compression stage in concrete. All three factors behave in the same manner as the loading rate is altered. The load at which the three factors occur increase as the loading rate is increased. This shows that the different material stages are also under the influence of inertia and damping during impact loadings. The effects of inertia seems to grow larger as the loading rate is increased.

Fig.3.46 shows the impact force at failure for different loading rates. The results from Section 3.3 are also included into this figure where the bending failure, bending to shear failure and punching shear failure regions are shown. The bending failure region can be classified as about 10*tf/msec* while the punching shear region is for loading rates higher that 80*tf/msec*. It can be noticed that a transition in the impact force at failure occurs between each failure region.

The deflection at failure is shown in Fig.3.47. Not much trend is noticed except that the deflection at failure tends to be larger as the loading rate is lower. The deflection for slabs under bending failure are small because the failure stage occurring in the slabs are different from the higher loading rate regions. During bending failure, the deflection reaches a maximum value before concrete crushing occurs.

Fig.3.48 shows the impulse versus loading rate relation. Impulse here is defined as the area enclosed by the impact load-time function for the calculations. From the figure, there is a tendency of the impulse being higher at the lower loading rates. The amount of impulse is the lowest at the medium range loading rate and then followed by a slight increase in impulse as the loading rate increases.

Fig.3.49 shows the amount of "*energy absorbed*" by the slab. The term "energy absorbed" mentioned here refers to the impulse being divided by the total energy. It gives a rough estimate of the ratio between the input load and the absorbed load. From the figure, it can be noticed that the values are higher at the lower loading rates. This can be attributed to the fact that the efficiency of the energy absorption process is higher under lower loading rates. For the lower loading rates, bending failure occurs first and the whole structure is capable of absorbing and dissipating the impulse received. On the other hand, for the higher loading rates where punching shear occurs, local failure is not efficient in re-distributing and also absorbing the amount of impulse from the impact load function.

3.4.4 EFFECTS OF CONCRETE TYPE [3.33, 3.34]

The combination of three types of concrete and two types of steel reinforcement are analytically studied with regard to improvement to impact performance. The three types of concrete are normal strength concrete (RC), high strength concrete (HRC) and steel fiber reinforced concrete (SFRC) while the steel reinforcements are the normal yield strength steel reinforcement (SD35) and high yield strength steel reinforcement (HT70). The amount of improvements in performance of each slab during impact loading are quantitatively and qualitatively studied by introducing the following indexes: loading rate, load at failure, deflection at failure, cracking load, impulse, total energy, local deformation index, volume displaced and impact failure mode.

Details of the analytical results of the six different slabs are given in Table 3.5 while the impact force-midspan deflection curves are shown in Fig.3.50. The initial part of the slab denotation indicates the concrete type while the latter part indicates the type of reinforcement used. The values in brackets in Table 3.5 indicate relative values to the RC-35 slab, which is the normal reinforced concrete slab.

Both SFRC slabs show the most improvement in the loading rate. A slight improvement is also expectable when HRC is applied. In general, the performance improvement can also be expected by replacing normal yield strength reinforcement with high yield strength reinforcement. The load at failure as well as the deflection at failure increase with the application of SFRC. Therefore, it can be concluded that application of SFRC improves ductility in the concrete slab tremendously. HRC also shows improvement in impact resistance but not as much as SFRC. Application of HRC contributes more to increasing the load at failure, but the deflection at failure is not much affected. High yield strength steel reinforcement is only effective when combined with RC or SFRC as no improvement is noticeable when used with HRC.

Improvement in the cracking load can be performed by applying HRC. But considering that the load at failure for HRC is lower than that of SFRC, it can be concluded that only the cracking load is effectively improved by HRC. But the decrease in stiffness after cracking is more evident than in SFRC. This phenomena is also noticeable in the immediate decrease in gradient of the curves in Fig.3.50 for the HRC slabs in comparison with SFRC slabs. Application of high yield strength steel reinforcement shows no distinct increase in the cracking load for HRC slabs.

Impulse is the amount of input enclosed by the impact force-time curve. Since the time to impact failure for all slabs are set at approximately 2 *millisec.*, the values for impulse are similar to that of the impact load at failure. Again, the SFRC slabs show the most improvement, followed by the HRC slabs. The total energy is defined as the area enclosed beneath the impact force-midspan deflection curve, i.e., indicating the amount of energy required for slab failure under a single impact. Subsequently, the total energy can be considered to be equivalent to the energy absorbed by the concrete slab. Generally, a change in load at failure and also in deflection at failure on the impact load-deflection relation would cause the amount of total energy to change too. Therefore, evaluating impact resistance by means of the total energy is actually a combination of two independent evaluations, that is evaluations according to the impact load at failure and also deflection at failure. Since the SFRC slabs show improvement in both the load at failure as well as the deflection at failure, the SFRC slabs are again the most efficient in terms of the total energy.

Local deformation is a main problem when dealing with impact loads, especially during impact loading with high loading rates. The curvature at failure (1/R) is introduced here to quantitatively study the effects of local deformation. Since the curvature is affected by the size of the maximum deflections at the loaded point, it is impossible to compare the curvature for different loading cases directly. An index of local deformation, as shown in the following equation,

$$\text{Index of local deformation} = \frac{\text{Curvature at failure } (1/R)}{\text{Deflection at failure}(\delta_u)} \qquad (3.53)$$

is introduced. The effects of deflections at the curvature at failure are eliminated by the above equation. A smaller index value indicates less local deformation at failure for the relative slab. The SFRC-35 slab followed by the RC-70 slab show the most improvement toward local deformation.

The volume displaced indicates the amount of volume displaced at failure in relation with the initial surface plane. The volume displaced gives an indication of the associated impact failure mode. The volume displaced is larger during bending failure, where displacement is spread all over the entire slab. On the other hand, during punching shear failure, the displacement is concentrated only at the center of the slab, and thus the volume displaced is comparatively smaller. The SFRC slabs show the most increase in volume displaced followed by HRC. But application of high yield strength steel reinforcement does not contribute much to improvement in volume displaced. The impact failure modes are improved by application of either HRC or SFRC.

3.4.5 EFFECTS OF FRP REINFORCEMENT [3.22]

The details of analytical results on normal strength concrete slabs with different types of FRP reinforcements are given in Table 3.6, while the corresponding impact force-midspan deflection curves are shown in Fig.3.51. In the denotations for the types of concrete slabs, the former part indicates the concrete type while the latter part shows the type of reinforcement used. Therefore, "RC-A1" would indicate a normal strength concrete slab with A1 reinforcement. The denotation "-35" would indicate the normal structural steel reinforcement while "-70" indicates high tensile strength steel reinforcement.

Comparing the impact load at failure shown in Table 3.6 and Fig.3.51, it is definite that the impact load at failure increases as the elastic modulus increases. As an example, the RC-A1 slab is 1.4 times that of the RC-B slab, while the RC-C1 slab is 0.6 times that of the RC-B slab. Similar trends are also noticeable in the cracking load. Furthermore, when the elastic modulus increases, the impact failure mode shows an improvement from the punching shear failure mode to the bending (flexural) mode. The impact load at failure tend to be equal when the elastic modulus is equivalent, but for the A series of slabs, the reinforcement with the smallest tensile strength, i.e., the RC-A3 slab has the lowest value. On the other hand, the RC-B slab has the lowest impact load at failure, in comparison to the RC-35 and RC-70 slabs.

The deformation mode at failure for the RC-A1, RC-B, RC-C1 and RC-D slabs are shown in Fig.3.52(a) ~ (d), respectively. In the RC-A1 slab, a total structural deformation in both the transverse and longitudinal directions is noticed, thus causing the failure mode to be bending failure. The effects of excitation of high modes of vibrations are noticeable in the RC-B and RC-C1 slabs, especially near the edges. The failure mode in these cases are the bending to punching shear failure mode, where the bending mode is dominant in the initial stages but as the failure progresses, the punching shear deformation mode becomes more prominent before final failure. In contrast, the deformation in the RC-D slab is mainly concentrated in the center of the slab from the initial stages, thus causing a local failure mode to be prominent. This phenomenon is mainly due to the dominant effects of shear deformations, thus causing the failure mode to be punching shear failure.

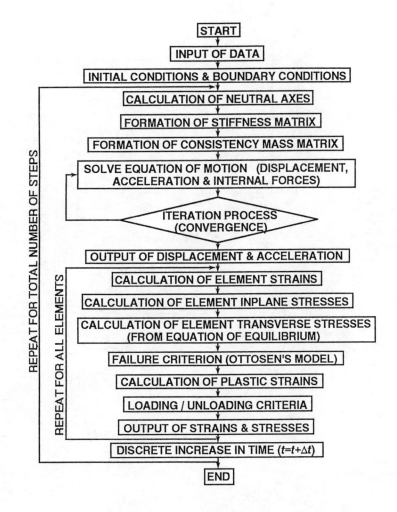

Fig. 3.1 Flow of analysis of RC slab structures

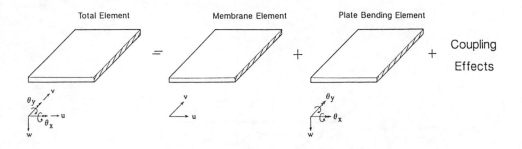

Fig. 3.2 Model of finite element

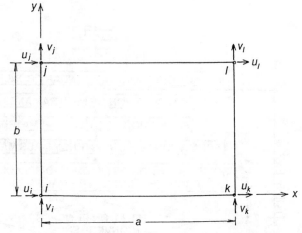

Fig. 3.3 Membrane element

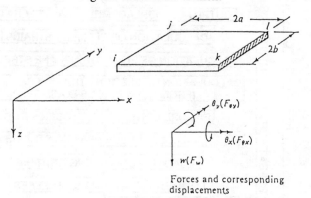

Forces and corresponding
displacements

Fig. 3.4 Plate bending element

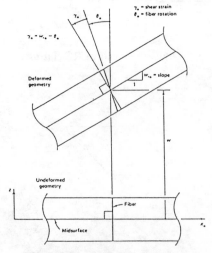

Fig. 3.5 Plate bending kinematics

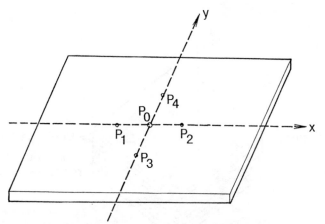

Fig. 3.6 Points for integrating element inplane stresses

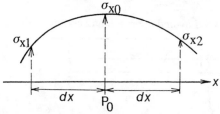

Fig. 3.7 Determination of partial derivatives

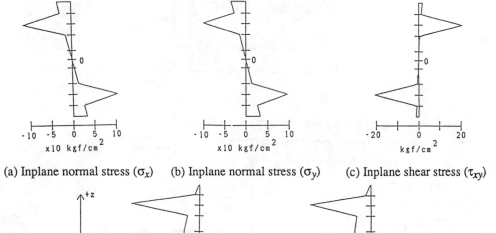

(a) Inplane normal stress (σ_x)   (b) Inplane normal stress (σ_y)   (c) Inplane shear stress (τ_{xy})

(d) Transverse shear stresses (τ_{xz})   (e) Transverse shear stresses (τ_{yz})

Fig. 3.8 Stress distribution in cross section of RC slab

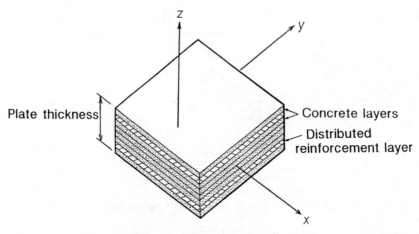

Fig. 3.9 Layered approach and coordinate axes

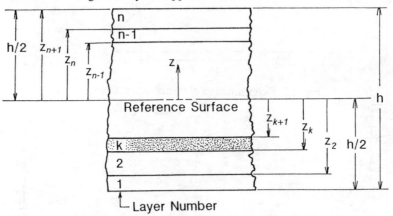

Fig. 3.10 Reference surface for layered approach

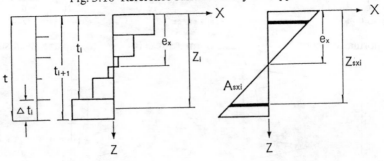

(a) Concrete   (b) Reinforcement

Fig. 3.11 Strain distribution in concrete and reinforcement cross-sections

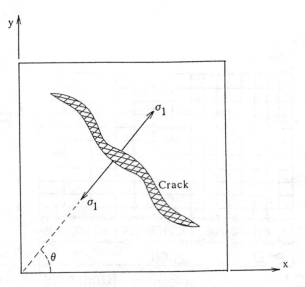

Fig. 3.12 Cracking in concrete element

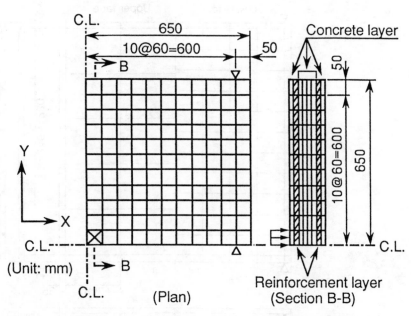

Fig. 3.13 Layered finite element meshes for RC slab (1/4 portion)

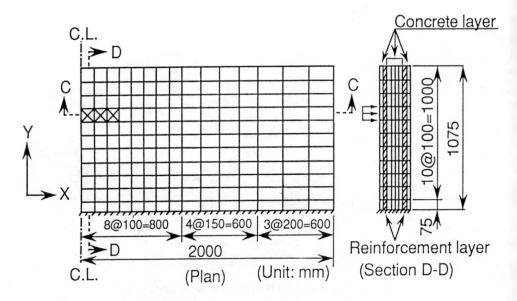

Fig. 3.14 Layered finite element meshes for RC handrail (1/2 portion)

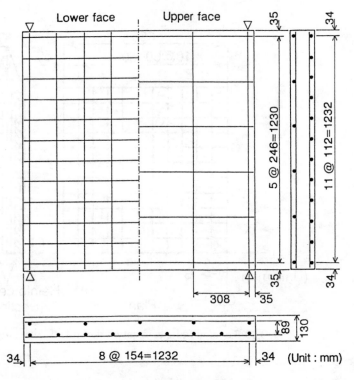

Fig. 3.15 Details of RC slab

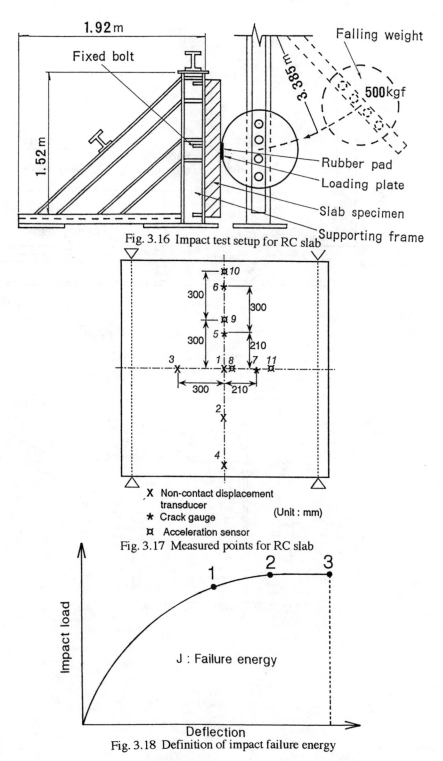

Fig. 3.16 Impact test setup for RC slab

Fig. 3.17 Measured points for RC slab

X Non-contact displacement transducer

* Crack gauge

¤ Acceleration sensor

(Unit : mm)

Fig. 3.18 Definition of impact failure energy

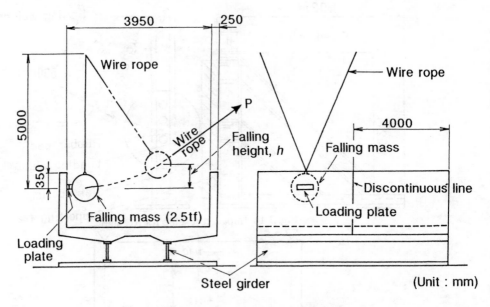

Fig. 3.19 Impact test setup for RC handrail

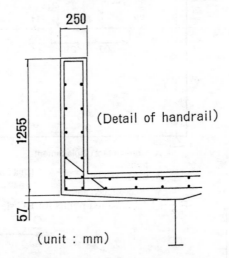

Fig. 3.20 Details of RC handrail

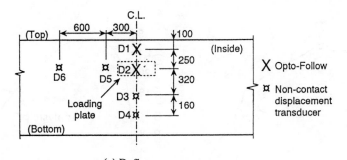

(a) Deflectometers

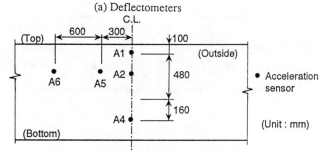

(b) Acceleration sensors

Fig. 3.21 Measured points for RC handrail

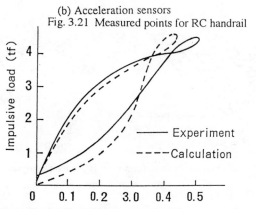

Fig. 3.22 Impact force-midspan deflection hysteresis curve for RC slab

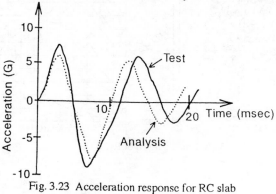

Fig. 3.23 Acceleration response for RC slab

164   Shock and Impact on Structures

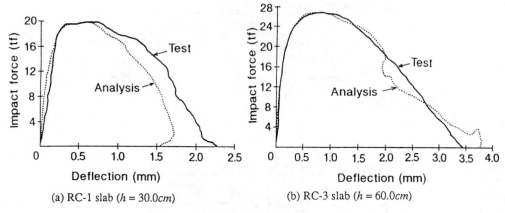

(a) RC-1 slab ($h = 30.0cm$) (b) RC-3 slab ($h = 60.0cm$)

Fig. 3.24 Impact force-midspan deflection curves for RC slabs

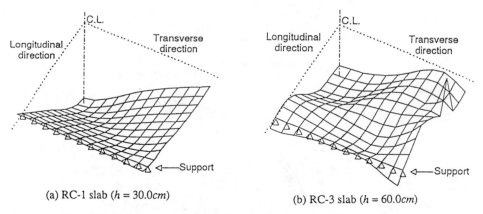

(a) RC-1 slab ($h = 30.0cm$) (b) RC-3 slab ($h = 60.0cm$)

Fig. 3.25 Deformation mode at failure for RC slabs (Analysis)

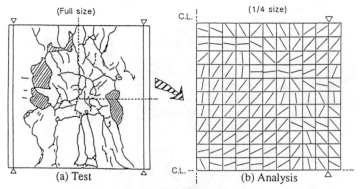

(a) Test (b) Analysis

Fig. 3.26 Crack pattern of RC-3 slab

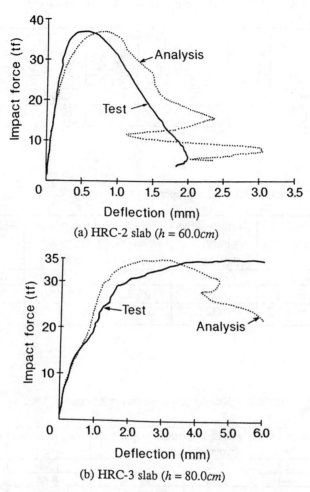

(a) HRC-2 slab ($h = 60.0cm$)

(b) HRC-3 slab ($h = 80.0cm$)

Fig. 3.27 Impact force-midspan deflection curves for HRC slabs

Table 3.1 Mechanical properties of FRP reinforcement

Abbrev.	Type	Cross section (cm2)	Yield point (kgf/cm2)	Yield strain (μ)	Tensile strength (kgf/cm2)	Tensile strain (μ)	Modulus of elasticity (kgf/cm2)	Poisson's ratio
SD35	Steel	1.267	3500	1667	4800	50000	2.10×10^6	0.300
Type 1	CFRP (HM35)	1.217	11070	7020	11070	7020	1.33×10^6	0.298
Type 2	CFRP (T700S)	1.216	21110	17000	21110	17000	1.01×10^6	0.327

Table 3.2 Mechanical properties of concrete

Slab	Compressive strength (kgf/cm^2)	Tensile strength (kgf/cm2)	Modulus of elasticity (kgf/cm2)	Poisson's ratio
RC-35	331.4	24.3	2.76×10^5	0.196
RC-1, RC-2	356.6	26.7	3.04×10^5	0.168
HRC-1, HRC-2	1032.6	38.7	3.75×10^5	0.217

Table 3.3 Results of test on concrete slabs

Slab	Height of fall (cm)	Loading rate (tf/msec)	Max. impact load (tf)	Max. deflection (mm)	Failure mode*
RC-35	65.0	14.32	28.64	12.44	PS
RC-1	40.0	8.35	22.54	8.64	PS
HRC-1	65.0	9.11	29.92	10.34	B
RC-2	65.0	8.62	24.99	13.91	PS
HRC-2	80.0	14.00	31.50	15.72	B

* PS = Punching shear, B = Bending

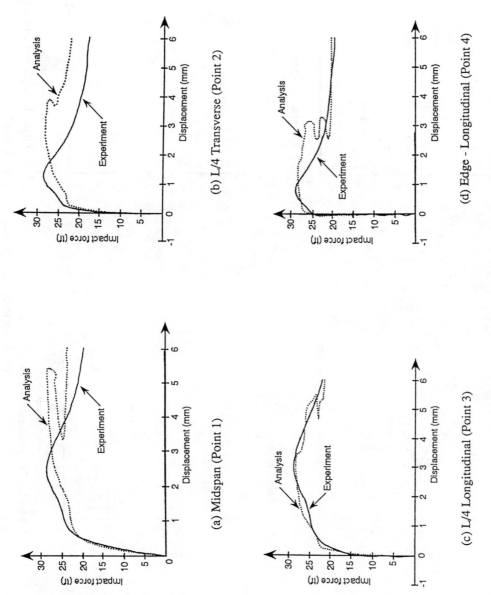

Fig. 3.28 Impact force-deflection curves for RC-35 slab

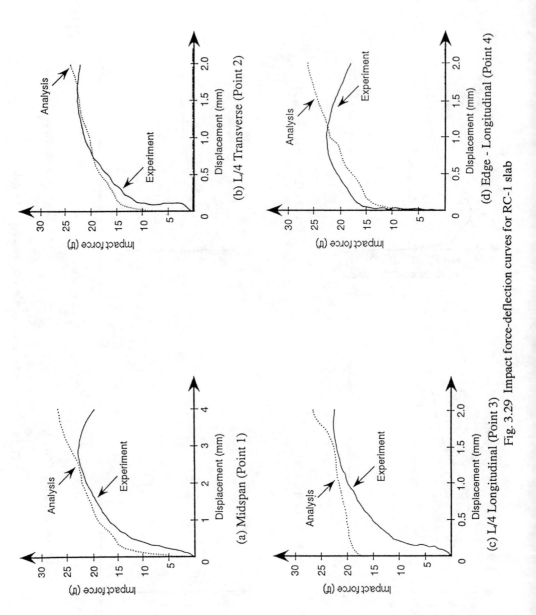

Fig. 3.29 Impact force-deflection curves for RC-1 slab

(a) Midspan (Point 1)

(b) L/4 Transverse (Point 2)

(c) L/4 Longitudinal (Point 3)

(d) Edge - Longitudinal (Point 4)

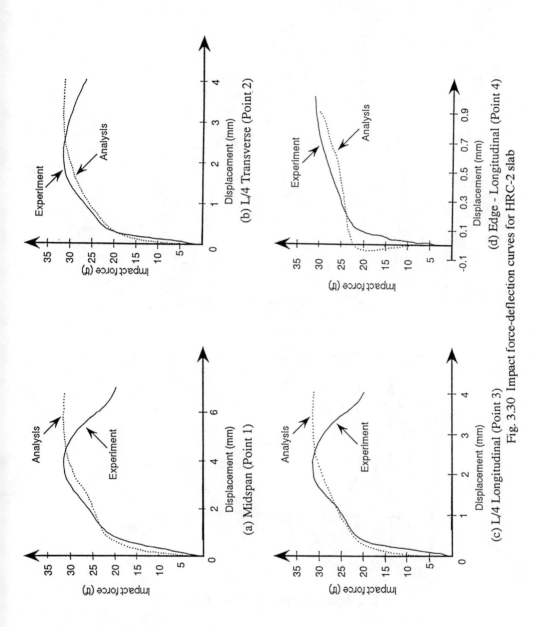

(a) Midspan (Point 1)

(b) L/4 Transverse (Point 2)

(c) L/4 Longitudinal (Point 3)

(d) Edge - Longitudinal (Point 4)

Fig. 3.30 Impact force-deflection curves for HRC-2 slab

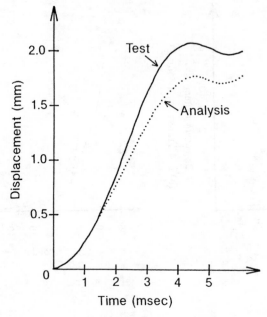

Fig. 3.31 Deflection-time relation for RC handrail

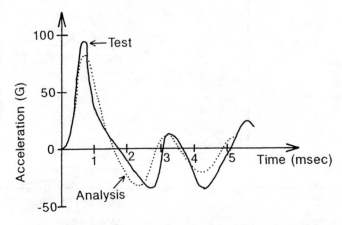

Fig. 3.32 Acceleration response for RC handrail

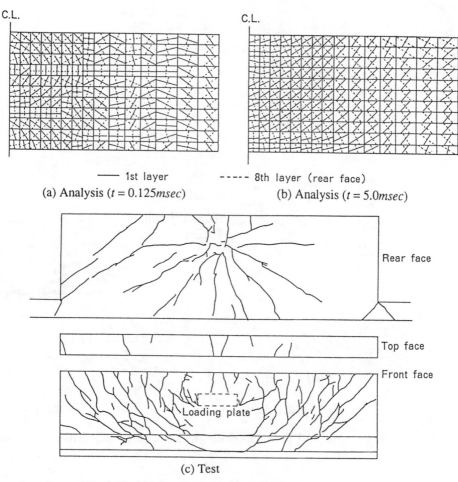

C.L.   C.L.

—— 1st layer   - - - - 8th layer (rear face)
(a) Analysis ($t = 0.125 msec$)   (b) Analysis ($t = 5.0 msec$)

Rear face

Top face

Front face

Loading plate

(c) Test

Fig. 3.33 Crack pattern at failure for RC handrail

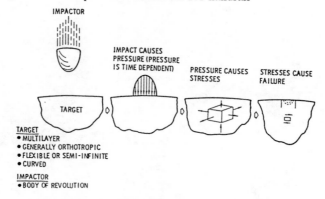

IMPACTOR

IMPACT CAUSES
PRESSURE (PRESSURE
IS TIME DEPENDENT)

PRESSURE CAUSES
STRESSES

STRESSES CAUSE
FAILURE

TARGET

TARGET
• MULTILAYER
• GENERALLY ORTHOTROPIC
• FLEXIBLE OR SEMI-INFINITE
• CURVED

IMPACTOR
• BODY OF REVOLUTION

Fig. 3.34 Failure mechanism under impact loading

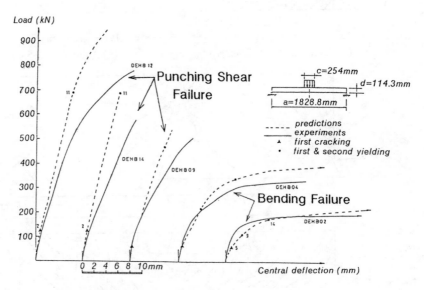

Fig. 3.35 Failure modes from midspan deflection curves

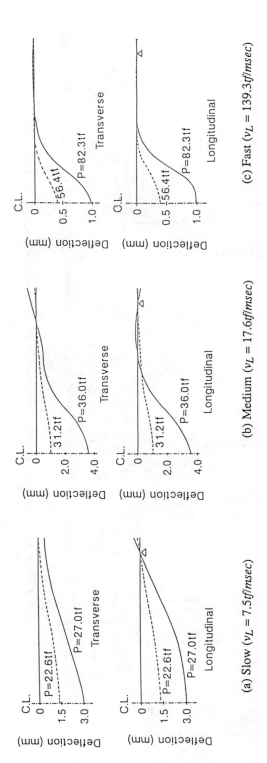

Fig. 3.36 Distribution of deflection for RC slabs under various loading rates

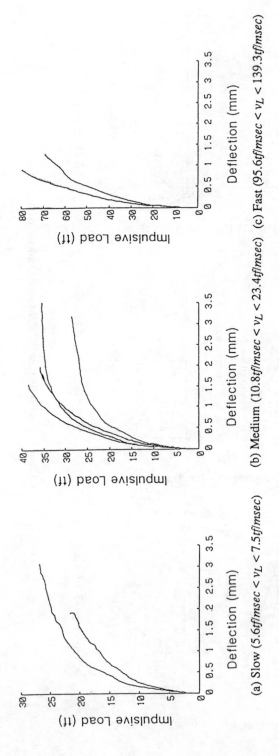

(a) Slow (5.6 *tf/msec* < v_L < 7.5 *tf/msec*)   (b) Medium (10.8 *tf/msec* < v_L < 23.4 *tf/msec*)   (c) Fast (95.6 *tf/msec* < v_L < 139.3 *tf/msec*)

Fig. 3.37 Impact force-midspan deflection curves for RC slabs under various loading rates

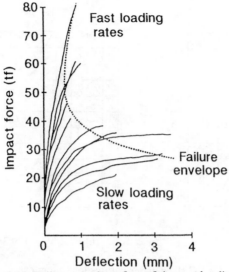

Fig. 3.38 Failure envelope for soft impact loadings of RC slabs

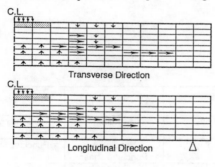

(a) Slow ($v_L = 7.5tf/msec$)

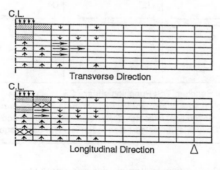

(b) Fast ($v_L = 95.6tf/msec$)

Circumferential crack Plastic (Compression)
Radial crack Reinforcement yielding

Fig. 3.39 Crack progression for RC slabs under various loading rates

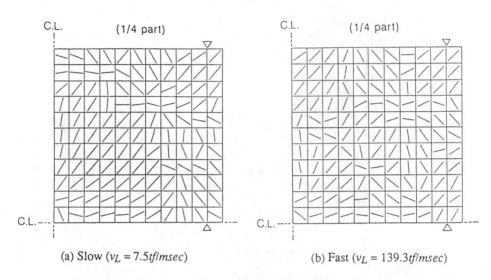

(a) Slow ($v_L = 7.5tf/msec$) (b) Fast ($v_L = 139.3tf/msec$)

Fig. 3.40 Crack pattern for RC slabs under various loading rates

Type of failure mode (Analysis)	B	B→PS (Mild)	B→PS (Severe)	PS
Deformation mode in primary stage	Bending	Bending	Bending	Punching shear
Deformation mode in secondary stage	Bending	Bending + Punching shear	Punching shear	Punching shear
Final failure mode	Bending	Punching shear	Punching shear	Punching shear
Deformation mode & failure condition	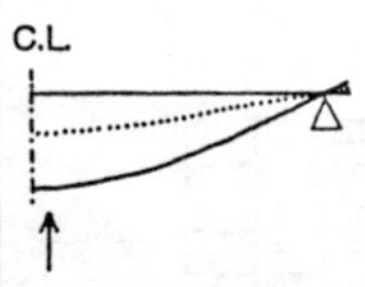	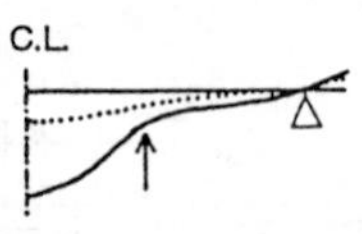	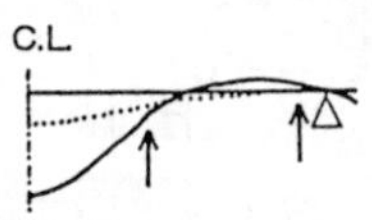	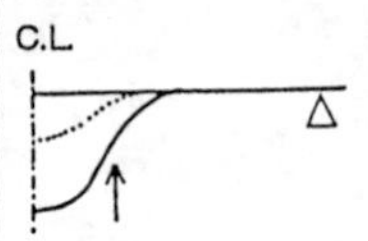

B: Bending, PS: Punching shear, C.L.: Center line
"Arrow" indicates typical area of element failure (Concrete crushing or reinforcement failure)
"Dotted line" indicates deformation in primary stage
"Solid line" indicates deformation in secondary stage

Fig. 3.41 Definition of impact failure modes and related failure conditions

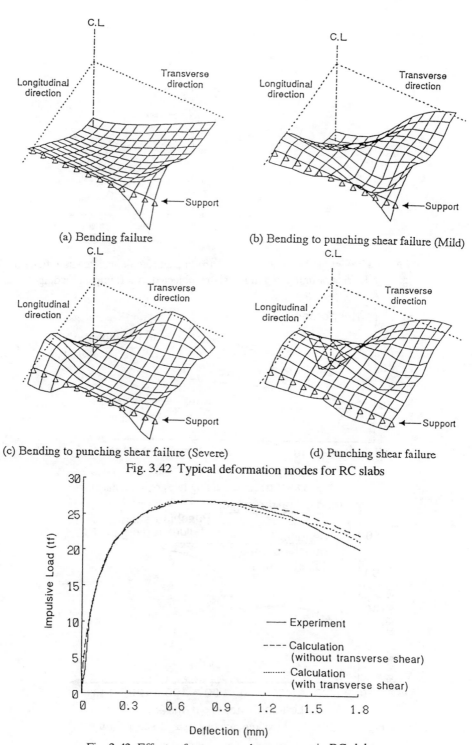

(a) Bending failure

(b) Bending to punching shear failure (Mild)

(c) Bending to punching shear failure (Severe)

(d) Punching shear failure

Fig. 3.42 Typical deformation modes for RC slabs

Fig. 3.43 Effects of transverse shear stresses in RC slabs

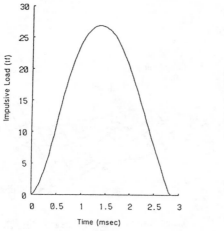

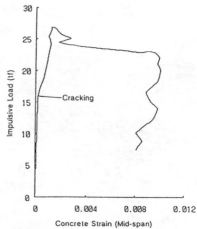

(a) Impact force-time function   (b) Impact force-concrete strain function

Fig. 3.44 Effects of unloading in RC slabs subjected to soft impact loading

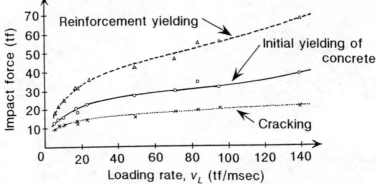

Fig. 3.45 Effects of loading rates on material states

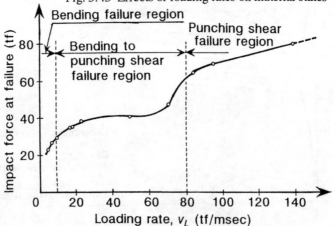

Fig. 3.46 Effects of loading rate on impact force at failure

Table 3.4 Results of analysis on concrete slabs under different loading rates

Loading rate, (tf/msec)	Load at failure, (tf)	Deflection at failure, (mm)	Elapsed time, (msec)	Cracking load, (tf)	Reinforcement yielding load, (tf)	Concrete* plasticity load, (tf)	Impulse, (kfg·sec)	Total energy, (kgf·cm)	Failure mode[1]
5.61	21.38	1.92	3.80	9.48	17.19	12.68	36.19	3019	B
7.50	26.98	3.07	3.60	11.36	21.43	14.50	50.56	6650	B
10.80	28.63	3.18	2.65	11.91	24.42	15.64	39.22	7664	B → P.S.
17.56	36.00	3.62	2.05	13.68	31.20	20.31	41.44	11,450	B → P.S.
17.90	35.80	1.98	2.00	12.66	30.16	17.87	32.74	5508	B → P.S.
23.40	38.60	1.56	1.65	14.38	35.49	22.17	32.03	4705	B → P.S.
49.51	40.85	0.70	0.83	15.90	42.03	26.48	15.44	2128	B → P.S.
71.12	46.23	0.59	0.65	18.61	46.23	29.49	13.81	7207	B → P.S.
83.72	64.88	0.78	0.78	19.26	55.17	33.96	27.45	8657	P.S.
95.58	69.29	0.88	0.73	20.21	55.41	31.16	23.93	6558	P.S.
139.25	79.89	0.89	0.57	21.64	67.58	38.00	20.01	5149	P.S.

*Plasticity in compression zone.
[1]B = bending, B → P.S. = bending to punching shear, P.S. = punching shear.
‡Yielding of reinforcement occurred after concrete crushing.
1 tf/msec = 9.807 × 10³ kN/s; 1 tf = 9.807 kN; 1 kgf·sec = 9.807 N·s; 1 kgf·cm = 9.807 × 10⁻² N·m.

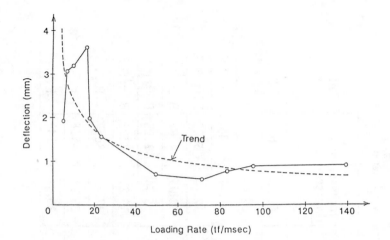

Fig. 3.47 Effects of loading rate on deflection at failure

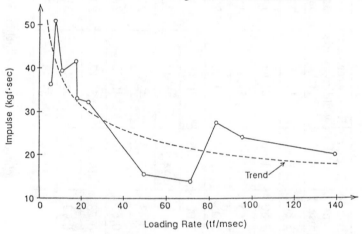

Fig. 3.48 Relation between impulse and loading rates

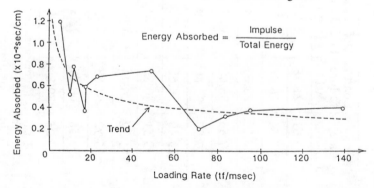

Fig. 3.49 Relation between energy absorbed and loading rates

Table 3.5 Results of analysis on concrete slabs with different concrete material

Slab	Loading Rate (tf/ms)	Load at Failure (tf)	Deflection at Failure (mm)	Cracking Load (tf)	Impulse (kgf·s)	Total Energy (kgf·cm)	LD[1] Index (x10^{-4}/cm^2)	Volume Displaced (cm^3)	Failure Mode[2]
RC-35	18.2 (1.00)	34.5 (1.00)	1.68 (1.00)	15.1 (1.00)	35.0 (1.00)	4797 (1.00)	10.70 (1.00)	169 (1.00)	B->PS
HRC-35	29.1 (1.60)	53.9 (1.56)	1.92 (1.14)	27.9 (1.85)	52.3 (1.49)	8223 (1.71)	9.98 (0.93)	255 (1.51)	B
SFRC-35	30.0 (1.65)	60.0 (1.74)	2.31 (1.38)	19.2 (1.27)	65.9 (1.88)	10990 (2.29)	4.67 (0.44)	338 (2.00)	B
RC-70	20.8 (1.14)	39.4 (1.14)	1.92 (1.14)	15.8 (1.05)	40.0 (1.14)	6228 (1.30)	7.18 (0.67)	190 (1.12)	B->PS
HRC-70	29.1 (1.60)	53.9 (1.56)	1.92 (1.14)	27.9 (1.85)	52.3 (1.49)	8223 (1.71)	9.96 (0.93)	254 (1.50)	B
SFRC-70	45.8 (2.52)	82.4 (2.39)	2.89 (1.72)	21.7 (1.44)	76.6 (2.19)	18750 (3.91)	8.37 (0.78)	349 (2.06)	B

[1] LD: Local Deformation (Transverse Direction)
[2] B: Bending, B->PS: Bending to Punching Shear

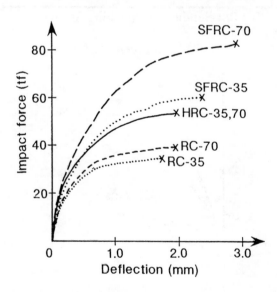

Fig. 3.50 Impact force-midspan deflection curves for concrete slabs

Table 3.6 Results of analysis on concrete slabs with different FRP reinforcements

Slab	Load at Failure (tf)	Deflection at Failure (mm)	Cracking Load (tf)	Failure Mode (1)	Failure Condition (2)
RC-A1	50.00	2.46	17.78	B	C
RC-A2	50.00	2.46	17.78	B	C
RC-A3	45.00	2.13	17.73	B	C
RC-B	33.76	1.84	16.00	B->PS	R
RC-70	39.44	1.92	15.76	B->PS	C
RC-35	34.51	1.68	15.13	B->PS	R
RC-C1	19.00	1.06	11.79	B->PS	R
RC-C2	18.00	1.15	11.86	B->PS	R
RC-C3	24.48	1.27	12.70	B->PS	R
RC-D	13.74	1.34	10.46	PS	R

(1) B: Bending, B->PS: Bending to Punching Shear,
    PS: Punching Shear
(2) C: Concrete crushing, R: Reinforcement failure

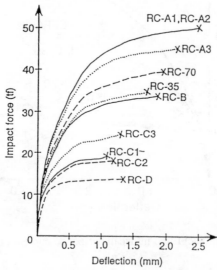

Fig. 3.51 Impact force-midspan deflection curves for concrete slabs (Various FRP reinforcements)

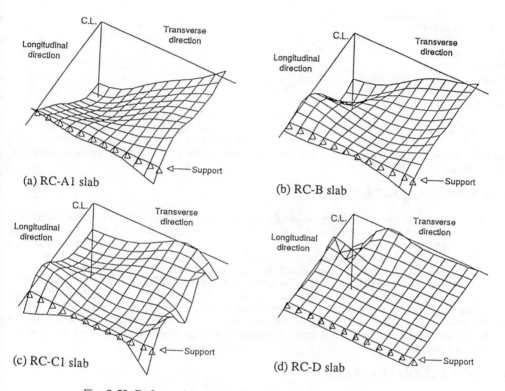

(a) RC-A1 slab

(b) RC-B slab

(c) RC-C1 slab

(d) RC-D slab

Fig. 3.52 Deformation modes for concrete slabs at failure

4. EVALUATION OF IMPACT RESISTANCE FOR REINFORCED CONCRETE SLABS

Evaluation of the degree of contribution of each type of concrete slab to impact resistance is necessary in the course of drafting design concepts for concrete slab structures that are subjectable to accidental impacts. Once each different type of concrete slab is quantitatively or qualitatively evaluated, an effective countermeasure for impact failure during accidental collisions is possible.

4.1 IMPACT RESISTANCE INDEXES

A method of evaluating the degree of impact resistance for concrete slabs is proposed based on the following indexes [4.1]:

(1) Impact load at failure,
(2) Deflection at failure,
(3) Cracking load,
(4) Total energy,
(5) Impulse,
(6) Index of local deformation,
(7) Volume displaced,
(8) Impact failure mode.

The effects of impact loading on reinforced concrete slabs based on these indexes are discussed. The types of concrete considered are normal strength concrete, high strength concrete, and steel fiber reinforced concrete, while the reinforcements are steel and FRP reinforcements, as shown in Table 2.4 and Fig.2.18.

4.1.1 IMPACT LOAD AT FAILURE

Fig.4.1 shows the impact load at failure for the different types of concrete slabs. Generally, introduction of steel fiber to reinforced concrete slab increases the impact load at failure for all combinations of reinforcement. Not much increase in impact load at failure is noticeable for both the HRC and SFRC slabs when reinforcement of the A and B series are applied. Therefore, it is considered most economical and practical to use the B series rather than the A series, as combinations with either HRC or SFRC. Furthermore, combining HRC or SFRC with the C series of reinforcement improves the performances of the concrete slabs to levels equivalent or even exceeding the impact performance of the normal reinforced concrete (RC-35) slab. In the case of the RC slabs, reinforcement with a high elastic modulus yields a larger impact load at failure. As for the D type of reinforcement, no distinct difference is noticeable in the impact load at failure for all types of concrete.

4.1.2 DEFLECTION AT FAILURE

The deflection at failure for concrete slabs is shown in Fig.4.2. Application of SFRC to reinforcements with modulus of elasticity equivalent or above that of reinforcing steel results in a remarkable improvement to deflection at failure. Comparing RC-A1 and RC-A2 to HRC-A1 and HRC-A2, respectively, the impact force at failure are comparatively similar, but the deflection at failure for RC-A1 and RC-A2 are remarkably larger. Therefore, there is no notable merit of applying high strength concrete when the reinforcements have a high modulus of elasticity. But in the case of steel reinforcement and C type of reinforcements, improvements in both impact force at failure and deflection at failure are evident for high strength concrete. It can be concluded that high strength concrete improves the impact performance of concrete slabs when reinforcement with a relatively low modulus of elasticity is applied.

4.1.3 CRACKING LOAD

The effects on cracking load for all types of combinations of slabs are shown in Fig.4.3. Overall, the HRC slabs have the highest cracking load, followed by the SFRC slabs and finally the RC slabs.

Therefore, the HRC slabs are the most efficient for preventing cracking in concrete structures during impact loading. Once again, the application of HRC and SFRC with the C series of reinforcements improve the impact performance of the slabs to levels equivalent to or above those of the RC-35 slab. The improvement in cracking load for each different type of concrete reaches a stable value when the ratio of elastic modulus is increased. For the SFRC and HRC slabs, no distinct improvement in cracking load is noticeable for combinations with reinforcement with a modulus of elasticity equivalent or above that of reinforcing steel. As for the normal strength concrete, distinct improvement in the cracking load is not notable for the A series of reinforcements.

4.1.4 TOTAL ENERGY

The total energy is defined here as the area enclosed beneath the impact load-midspan deflection curve, i.e., indicating the amount of energy required for slab failure under a single impact loading. Subsequently, the total energy can be considered to be equivalent to the energy absorbed by the concrete slab. Generally, a change in impact load at failure and also in deflection at failure on the impact load-deflection relation would cause the amount of total energy to change too. Therefore, evaluating impact resistance by means of the total energy is actually a combination of two independent evaluations, that is evaluations according to the impact load at failure and also deflection at failure.

The results of total energy for the different types of slabs are shown in Fig.4.4. The total energy for RC and SFRC slabs increase as reinforcement with a larger elastic modulus is applied. A notable increase in total energy is evident in the case of SFRC being combined with the A series of reinforcements. Application of HRC or SFRC to the C series of reinforcements produces results that are above those of the normal RC-35 slab. But for reinforcements with very low elastic modulus, such as the D type of reinforcement, no distinct differences are present for all types of concrete. Generally, application of SFRC together with reinforcements having a modulus of elasticity similar or above those of the steel reinforcement would produce distinct improvements to the total energy.

4.1.5 IMPULSE

Impulse is defined as the area enclosed beneath the impact force-time curve. Impulse for the different types of concrete slabs are shown in Fig.4.5. Since the time at failure is set at approximately 2 *msec* for all types of slabs, in correspondence to test results, the amount of impulse is proportional to the impact force at failure. Therefore, the results in Fig.4.5 are similar to results shown in Fig.4.1. Consequently, combination of SFRC and A type of reinforcements show a remarkable increase in impulse. As for the D type of reinforcement, there is no distinct improvement in amount of impulse for all types of concrete. Therefore, it is more economical to use normal strength concrete in combination with the D reinforcement.

4.1.6 CURVATURE AT FAILURE

Local deformation during impact loading can be quantitatively evaluated based on the curvature at failure. In order to erase the effects of maximum deflection for different types of slabs, the index of local deformation, defined in Eq.(3.53), is employed here. The index of local deformation in the transverse (span) direction for concrete slabs are given in Fig.4.6. Generally, concrete slabs with the D reinforcement show very large indexes of local deformation, indicating local type of failure. No distinct difference in the index of local deformation for other types of concrete slabs are noticeable. The major cause is because the effects of higher modes of vibration are not expressed in the following index. Fig.4.7 shows the distribution of deflection for the bending and bending to punching shear failure modes. Depending on the value of Δx for determining the curvature at failure, the effects of high modes of vibration are not exactly expressed in this particular case ($\Delta x = 18.0 cm$). A better representation of local deformation can be obtained by selecting the most suitable value of Δx.

4.1.7 VOLUME DISPLACED

The volume displaced indicates the amount of volume displaced at failure in relation with the initial surface plane. It is calculated by integration of deflection for the entire surface of the structure. The volume displaced gives an indication of the associated impact failure mode. The volume displaced is larger during bending failure, where displacement is spread all over the entire slab. On the other hand, during punching shear failure, the displacement is concentrated only at the center of the slab, and thus the volume displaced is comparatively smaller. It is considered that the volume displaced gives a better representation of the amount of energy absorbed, in relation to total energy, as the entire deformation of the structure is taken into consideration. The total energy is valid for bending type of failure modes, where the distribution of deformation between supports and midspan can be considered to be linear. But in the case of local or punching shear failure, the distribution of deformation is highly localized, and total energy in this situation does not reflect the actual amount of energy that has been absorbed by the structure.

The results for the volume displaced in the concrete slabs are shown in Fig.4.8. On the whole, a combination of SFRC with the A, B and C series of reinforcements results in distinct improvements to the volume displaced. As for the application of HRC, improvements to the volume displaced are evident for reinforcements having elastic moduli smaller than the A series. Similarly, application of either HRC or SFRC to the C series of reinforcements indicate an equal or better performance in comparison with the RC-35 slab. As for the D reinforcement, no differences are noticeable for either types of concrete.

4.1.8 IMPACT FAILURE MODE

The impact failure modes for various types of concrete slabs are shown in Fig.4.9. Generally, reinforcements with a higher modulus of elasticity tend to show an improved impact failure mode. The A series of reinforcements indicate bending failure when applied with any type of concrete material. As for the B type of reinforcement and also the steel reinforcements, improvements in the failure mode in relation to the RC slab, from the bending to punching shear failure mode to a bending failure mode, is possible by applying either HRC or SFRC. The failure mode for the C2 and C3 reinforcements are improvable by a combination with SFRC but no distinct improvements are possible for the C1 reinforcement. Therefore, it is considered that reinforcements with a plastic region, as in the C2 and C3 reinforcements, would result in a better performance to impact loads. The punching shear failure mode for D reinforcement can be relatively improved to the bending to punching shear failure mode by a combination with either HRC or SFRC.

4.1.9 LOADING RATE

The effects of loading rate on four different types of concrete slabs are considered here, namely normal strength reinforced concrete (RC) slab, light weight aggregate reinforced concrete (LRC) slab, steel fiber reinforced concrete (SFRC) slab and high strength reinforced concrete (HRC) with a combination of SD-35 steel reinforcement bar. The effects of loading rate on the impact load at failure, deflection at failure and total energy will be considered [4.2, 4.3].

Fig.4.10 shows the impact load at failure-loading rate relation. Under an increasing loading rate, the impact force at failure shows an increase mainly because of inertia. This phenomenon can be noticed clearly in the figure. The HRC slab shows a large impact force at failure when compared to the other slabs, with an increase of about 20% compared to the RC slab. Not much difference is seen in the RC and SFRC slabs, but the LRC slab shows a low failure load on the whole (about - 10% compared to RC).

Fig.4.11 shows the relation between loading rate and midspan deflection at failure. It can be considered as a rough estimate of the deformation capability of the slabs. A decrease in deformation is noticed as the loading rate increases. The decrease or drop in deformation capability is quite small in the SFRC slab, having almost a stable value throughout. The LRC slab shows a large deformation capability, especially in the slow loading rate regions while the HRC slab shows small deflections on the whole.

Fig.4.12 shows the total energy in relation with loading rate. Since total energy is defined as the amount of energy required for slab failure under a single impact, it can be considered as a combination of the other two indexes, namely the failure load and the deformation capability. Thus, evaluations by means of this index could prove to be more effective. Total energy increases as the loading rate is increased. The SFRC slab seems to show a larger increase in index as the loading rate increases. The HRC slab shows high values in the slower loading rate regions but as the loading rate increases, the SFRC slab has a larger index value. The LRC slab shows small energy absorption properties on the whole.

4.2 DESIGN CONCEPTS FOR REINFORCED CONCRETE STRUCTURES

Impact design methods of most structures are carried out by adopting an impact factor in the static design method. The dynamic forces during impact are converted into a static force of equal magnitude and treated in much the same way as other static loads. This static design method would not adequately describe an impact phenomenon and thus have only a limited amount of practical applicability. During impacts, an excitation of not only the first mode but also higher modes of vibration can be expected. Structures designed using the equivalent dynamic force would be able to withstand bending but not the shear, hence bringing about punching shear or concrete scabbing. These factors can only be totally considered if a dynamic design approach is adopted.

4.2.1 ULTIMATE STATES OF CONCRETE STRUCTURES

The ultimate states of a concrete structure are necessary during the design process. Since impact loading on concrete structures has a very low occurence probability, the normal case would be to design the structure according to the ultimate limit states. A serviceability limit state would result in an uneconomical and conservative design. The ultimate states for different types of concrete structures are totally different. Some examples of ideal ultimate limit states for concrete structures in the field of civil and structural engineering are listed in Table 4.1. The ultimate limit state for rock sheds under impact loading can be set as structural failure, either in the bending or shear failure modes. But in the case of concrete handrails, the ultimate limit state would also be structural failure. But the resultant impact failure mode would be a very important factor. Another limit state for concrete handrails, especially in overhead expressways or in multi-storey intersections (crossings), is the prevention of scabbing at the rear face, as concrete scabbing would likely cause a secondary disaster when falling onto a different traffic lane below. Concrete scabbing can be effectively reduced or even prevented by allowing a bending failure mode to occur. The formation of shear cones during shear failure would cause concrete scabbing to occur easily. Therefore, for the design of concrete handrails, the bending failure mode would be important.

A different ultimate state would be for the design of nuclear power plants and its related facilities, where cracking in the concrete structure could cause leakage of radioactive materials into the atmosphere. In certain cases, where the structure is not highly radioactive, then prevention of perforation would be a more economical ultimate state. Structural integrity would be a major ultimate state when considering a structural system. For example, in the design of marine offshore structures or gravity platforms, or even the design of buildings, failure of structural elements would be the limit state, but the total structure should be intact after the impact. A loss in the bearing capacity of a structural element could cause the entire structure to collapse.

The amount of permanent deformation is also the ultimate limit states in certain cases. For example, the permanent deformation in a bridge pier should be limited, or else it would cause damage or even collapse of the super-structure.

The ultimate states for reinforced concrete handrail will be discussed a little more in detail here and followed by a case study for design in the following Section. An ideal design procedure of concrete handrails for expressways is relatively difficult. An ideal handrail should be able to withstand the impact from a colliding vehicle. The handrail should not act as a solid barrier to stop the collision but more as a flexible wall that is capable of absorbing most of the impact collision energy. Therefore, it is necessary to design concrete handrails to fail under bending, as energy absorption is better during ductile type of failure. During the collision of a light vehicle or when the impact momentum is relatively small, the handrail should act as a rigid barrier and allow the deformation of the vehicle itself to absorb most of the impact energy. When the momentum of the collision is large, or when the collision speed is large, then the handrail should deform and absorb most of the impact energy, especially in the bending mode, and possibly let the vehicle not to have relatively large deformations. Therefore, the life or lives of occupants in the vehicle would not be highly endangered. In the case of collisions by large trucks or trailers, the structure should absorb the energy but not allow scabbing of concrete to occur at the rear face.

4.2.2 DESIGN CONCEPTS FOR REINFORCED CONCRETE STRUCTURES

The energy criterion would be the most efficient method of designing concrete structures under impact loads. The external energy from the impacting body (impactor) should be absorbed by the entire structure. The energy transformation process during impact loading of concrete structures is illustrated in Fig.4.13[4.4]. The main part of the kinetic energy in an impacting body is transmitted to the concrete structure during impact collision. The energy transmitted to the concrete structure is then converted into the kinetic energy of the concrete structure and also the energy absorbed by the structure. The energy absorbed by the structure is converted into irrecoverable energy and also strain energy, which is recoverable. Formation of cracks and also fracture zones , friction, damping and etc. are the main source of the irrecoverable energy. When there is vibration in the concrete structure, then part of the strain energy is then re-transmitted as kinetic energy back to the concrete structure, as indicated in the figure by broken lines.

Besides the energy criterion, most design specifications for structures under impact loading specify the impact load-time function as the design impact load. Fig.4.14 shows the idealized design impact load for impact of *Phantom* fighter aircraft into concrete nuclear reactors at a speed of 215km/sec [4.5]. This design impact load is used for design of certain concrete nuclear reactors in Germany. In such a case, an impact load criterion must be applied for design of the structure. Therefore, a dynamic design method, based on both the energy and load criteria, is proposed here. Fig.4.15 shows the proposed design method for concrete structures.

Bold lines in the flow chart indicate the safety provision according to energy criterion while the broken lines indicate safety provisions according to load criterion. The flow of the proposed design method can be explained as follows, where the numbers indicated correspond to the sequence of numbering in the flow:

(1) Outline of type of structure - The type of structure to be designed is selected, i.e., concrete handrail, concrete pier, etc.

(2) Is design impact load specified? - If the design impact load is specified for the particular case, then further considerations regarding the impact load is not necessary. If the design impact load is not specified, then considerations of the type and characteristics of the impact load function must be predicted based on a numerical or empirical procedure.

(3) Preliminary study and survey of possible impact collisions - The type of impacting body that could possibly impact on the structure and also the possible collision speeds and also collision angles have to be surveyed.

(4) Determination of design impact conditions - Based on the results obtained from the survey in Step (3), the design impact conditions (type of impacting body, collision angle, collision speed, etc.) are selected based on statistical considerations.

(5) Is safety factor required? - It is considered that the safety factor can be calculated by applying the energy criterion. In the case of load criterion , only a straight-forward procedure is possible, where the structure is checked for failure under the specified impact load condition.

(6) Calculation of design impact load - The impact design load is predicted based on numerical models. In this dissertation, application of the multi-mass model is proposed for predicting the design impact load function.

(7) Selection of type of structure - Further details of the concrete structure is selected. By considering the magnitude of the specified design impact load or specified design impact collision, a structure that can effectively withstand the impact collision is selected, i.e., reinforcedconcretestructure or prestressed concrete structure, simple supports or fixed supports, etc.

(8) Results of design for static loads - At present, most structures are not designed specifically for impact loading conditions. The structure is usually designed to withstand the necessary design loads, such as dead load, live load, earthquake load, etc. Once the dimensions of the structure have been determined based on the specified static design codes to withstand the normal loading conditions, the impact loading condition is then applied to the structure.

(9) Setting of structural dimensions - If design of the structure to resist normal loading conditions in

Step (8) has been performed, then the structural dimensions are fixed. But if no preliminary design under normal static loading conditions have been performed, then the structural dimensions are selected.

(10) Dynamic structural analysis - Dynamic structural analysis is then performed based on the specified design impact load or design impact condition. In the context of this dissertation, the dynamic nonlinear finite element analysis proposed earlier is applied. The "linked" procedure is applied for the safety provision by energy criterion while the normal finite element procedure is applied for the safety provision by load criterion.

(11) Is energy criterion satisfied? - Safety provision based on the energy criterion is applied. Details are given in the following section.

(12) Is failure condition exceeded? - Safety provision based on the load criterion is applied. Details are given in the following section.

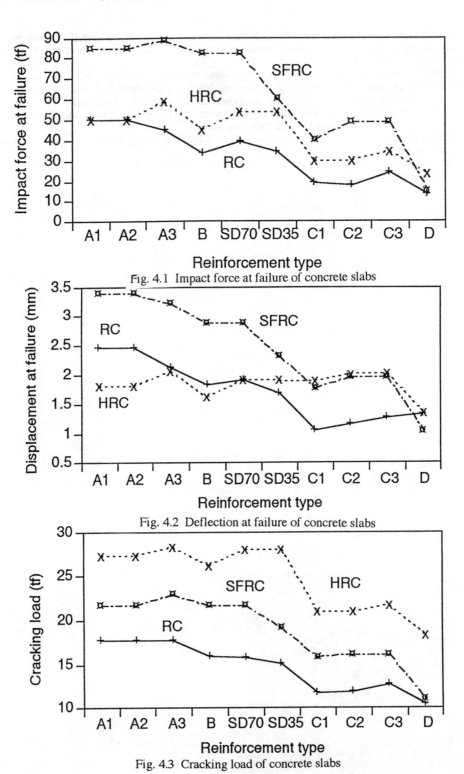

Fig. 4.1 Impact force at failure of concrete slabs

Fig. 4.2 Deflection at failure of concrete slabs

Fig. 4.3 Cracking load of concrete slabs

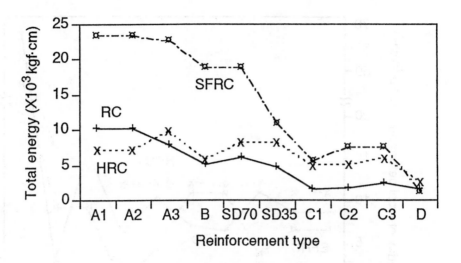

Fig 4.4 Total energy of concrete slabs

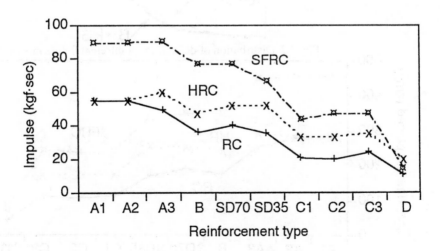

Fig. 4.5 Impulse of concrete slabs

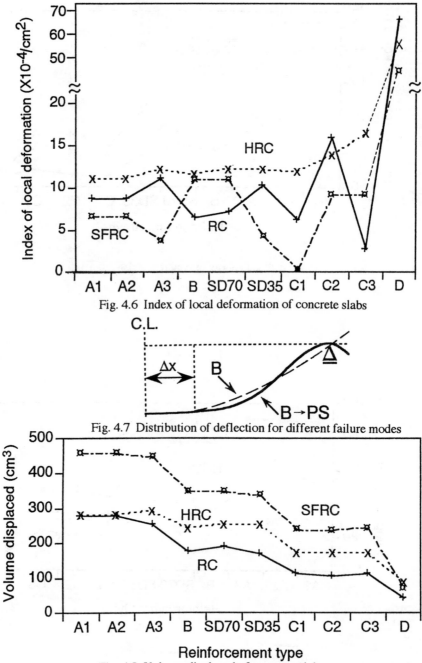

Fig. 4.6 Index of local deformation of concrete slabs

Fig. 4.7 Distribution of deflection for different failure modes

Fig. 4.8 Volume displaced of concrete slabs

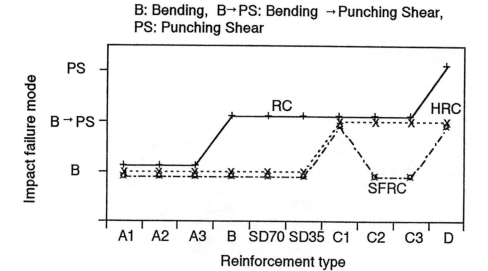

Fig. 4.9 Impact failure mode of concrete slabs

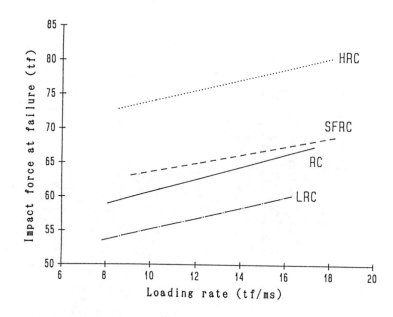

Fig. 4.10 Effects of loading rate on impact load at failure (Slabs)

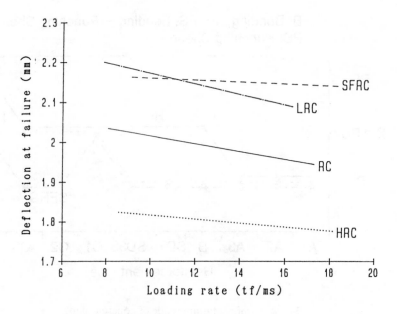

Fig. 4.11 Effects of loading rate on deflection at failure (Slabs)

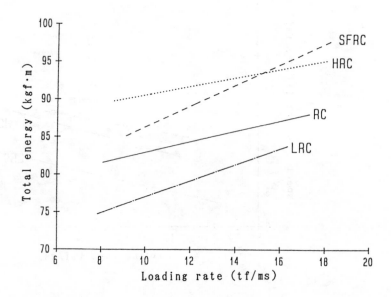

Fig. 4.12 Effects of loading rate on total energy (Slabs)

Table 4.1 Ultimate states of concrete structures under impact loads

Load characteristic	Impactor	Target / Structure	Ideal ultimate limit states
Soft	Vehicle	Handrail / Barrier	Failure in structural element. Energy is absorbed by flexural deformation of structure. Punching shear and concrete scabbing should be prevented.
		Building	Energy is absorbed by failure in structural element. Collapse of entire structural system should be prevented by allowing hinges to form at beam sections.
		Bridge pier	Energy is absorbed by failure in structural element. But the stability of entire structure should be ensured, especially in the supporting forces. Furthermore, deformations that would cause movements to the upper structure should be checked.
	Ship	Bridge girder	Failure in structural element. Large deformations would cause collapse of the girders from piers. Adequate flexibility should be allowed to prevent collapse from supporting piers.
		Bridge pier	Energy is absorbed by failure in structural element. But the stability of entire structure should be ensured, especially in the supporting forces. Furthermore, deformations that would cause movements to the upper structure should be checked. At deep water levels, the piers should be designed to be rigid, i.e., collision energy should be absorbed by deformation of ship.
		Offshore structure Marine structure Gravity platform	Energy is absorbed by failure in structural element. But the stability of entire structural stability should be ensured.
	Aircraft	Nuclear power plant Important structure Protective shelter	Single layered structure: Cracks should not be allowed for structures of extreme importance. In certain cases, cracking should be allowed but penetration and scabbing should be prevented at all cause. Double protection structure: Penetration and scabbing are allowable in the secondary structure. Cracks should not be allowed in the primary structure for extremely important structures. In certain cases, cracking should be allowed but penetration and scabbing should be prevented at all cause in the primary structure.
	Rock	Rock / Snow shed	Failure in structural element.
Hard	Explosion	Protective shelter	Single structure: Cracks should not be allowed for structures of extreme importance. In certain cases, cracking should be allowed but penetration and scabbing should be prevented at all cause. Double protection structure: Penetration and scabbing are allowable in the secondary structure. Cracks should not be allowed in the primary structure for extremely important structures. In certain cases, cracking should be allowed but penetration and scabbing should be prevented at all cause in the primary structure.

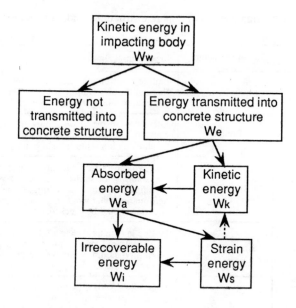

Fig. 4.13 Energy transformation process of impact collision

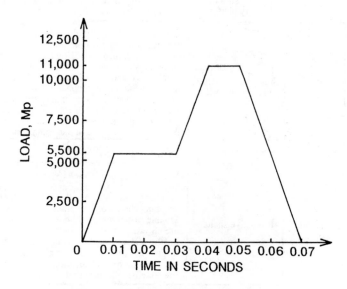

Fig. 4.14 Aircraft impact loading function

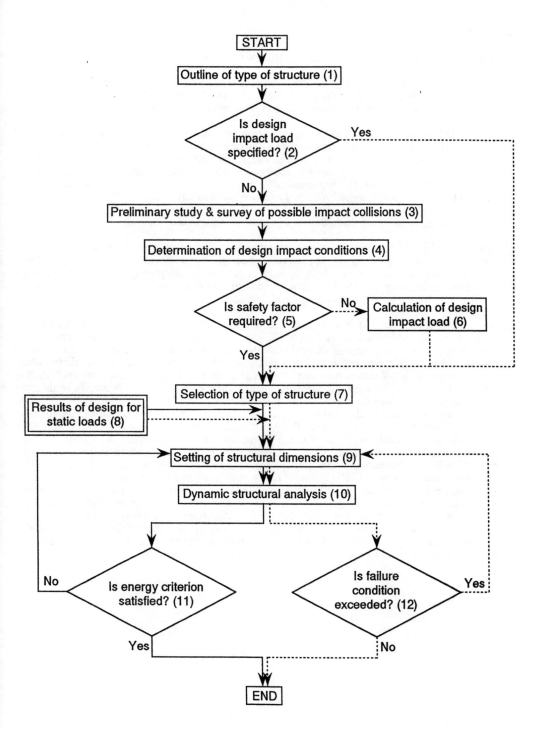

Fig. 4.15 Flow of dynamic design method for structures under impact load

5. CONCLUSIONS

In this paper, the two main objectives are satisfied as follows;

(1) Development of dynamic analytical procedure for predicting the ultimate behaviors as well as the impact failure modes during impact loading of concrete structures is performed by applying a layered finite element procedure which is developed for analysis of reinforced concrete slab structures.
(2) Proposal of quantitative and qualitative evaluation procedures for concrete structures under soft impact loads is performed by numerical analysis on different types of reinforced concrete slab structures. Indexes such as impact load at failure, deflection at failure, total energy, failure region, cracking load, impulse, index of local deformation, volume displaced and impact failure modes are proposed. And also, a dynamic design procedure for concrete structures subjected to impact loads is proposed based on energy criterion and load criterion.

The main conclusions obtained in this paper can be summarized as follows:

(1) The layered finite element procedure, together with considerations of the transverse shear stresses, is capable of giving good predictions of the ultimate behaviors, such as material states, impact failure load, deflection at failure, crack pattern, and crack propagation, of concrete slabs structures under impact loads. Furthermore, the impact failure modes of concrete slabs can be predicted based on the deformation mode, impact load-deflection curves, and the crack behaviors.
(2) The impact failure modes for reinforced concrete slab structures under soft impact loading are found to be related to the loading rate of the impact force-time function. Three distinct failure modes are analytically identified and classified as; bending failure, intermediate failure (bending to punching shear failure), and punching shear failure. In the intermediate failure mode, the bending deformation mode is dominant at the initial stages, but the effects of higher modes of vibrations causing the punching shear failure mode becomes dominant in the later part of impact. The final failure mode in this situation is punching shear failure.
(3) Indexes such as impact load at failure, deflection at failure, cracking load, total energy, impulse, index of local deformation, volume displaced and impact failure modes are effective for quantitatively and qualitatively evaluating the impact resistance properties of concrete slabs under impact loads. Evaluations show that reinforcements with a high modulus of elasticity show a better performance under impact loading. Furthermore, application of steel fiber reinforced concrete into a similar concrete slab would result in a high impact resistant structure. Cracking load in a structure can be effectively reduced by applying high strength concrete. Consequently, the impact failure modes in reinforced concrete slabs can be improved from the material point of view, by proper combinations of reinforcement and concrete materials.
(4) A dynamic design procedure that is capable of considering the impact failure modes and also the safety factor is proposed based on an energy criterion. Furthermore, to allow application of specified design impact loads in design, a load criterion is also proposed.

REFERENCES

1.1 Bangash, M. Y. H., *Concrete and Concrete Structures: Numerical Modelling and Applications*, Elsevier Applied Science, London (UK), 1989, 668 pp.
1.2 Lin, T. D., "Concrete for Lunar Base Construction," *Concrete International*, July 1987, pp. 48-53.
1.3 Kar, A. K., "Impactive Effects of Tornado Missiles and Aircraft," *Journal of the Structural Division*, ASCE, Vol. 105, No. ST11, Nov. 1979, pp. 2243-2260.
1.4 Davies, G. A. O., (ed.), *Structural Impact and Crashworthiness (Vol. 1 - Keynote Lectures)*, Elsevier Applied Science Publishers, London (UK), 1984, 242 pp.
1.5 Morton, J., (ed.), *Structural Impact and Crashworthiness (Vol. 2 - Conference Papers)*, Elsevier Applied Science Publishers, London (UK), 1984, 555 pp.
1.6 Jones, N., and Wierzbicki, T., (eds.), *Structural Crashworthiness*, Butterworths, London (UK), 1983.
1.7 Wierzbicki, T., and Jones, N., (eds.), *Structural Failure*, John Wiley & Sons, New York, 1989, 551 pp.
1.8 *Concrete Structures Under Impact and Impulsive Loading - Introductory Report*, RILEM, CEB, IABSE, IASS Interassociation Symposium, Berlin (West), June 2-4, 1982, 147 pp.
1.9 *Concrete Structures Under Impact and Impulsive Loading - Proceedings*, RILEM, CEB, IABSE, IASS Interassociation Symposium, Berlin (West), June 2-4, 1982, 656 pp.
1.10 *Concrete Structures Under Impact and Impulsive Loading - Final Report*, RILEM, CEB, IABSE, IASS Interassociation Symposium, Berlin (West), June 2-4, 1982, 160 pp.
1.11 Bulson, P. S., (ed.), *Structures Under Shock and Impact*, Computational Mechanics Publications, Southampton (UK), 1989, 533 pp.
1.12 Bulson, P. S., (ed.), *Structures Under Shock and Impact II*, Computational Mechanics Publications, Southampton (UK), 1992, 674 pp.
1.13 Maekawa, I., (ed.), *Proceedings of the International Symposium on Impact Engineering* - Vol. I & II, Sendai (Japan), Nov. 2-4, 1992, 775 pp.
1.14 *International Workshop on Blast Resistant Structures - Proceedings*, Beijing (P. R. China), Oct. 14-16, 1992, 243 pp.
1.15 *Proceedings of Symposium on Impact Problems Related with Falling Rocks*, Technical Committee on Impact Problems, JSCE, Tokyo, Mar. 22-23, 1991, 156 pp.
1.16 Hughes, G., and Beeby, A.W., "Investigation of the Effect of Impact Loading on Concrete Beams," *The Structural Engineer*, Vol. 60B, No. 3, Sept. 1982, pp. 45-52.
1.17 Miyamoto, A., King, M.W., and Fujii, M., "Nonlinear Dynamic Analysis of Reinforced Concrete Slabs Under Impulsive Loads," *Journal of the American Concrete Institute*, Vol. 88, No. 4, July-Aug. 1991, pp.411-419.
1.18 Brandes, K., "Behaviour of Critical Regions Under Soft Missile Impact and Impulsive Loading," RILEM, CEB, IABSE, IASS Interassociation Symposium: *Concrete Structures Under Impact and Impulsive Loading - Introductory Report*, Berlin (BAM), June 2-4, 1982, pp.89-112.

2.1 Bazant, Z. P., and Parameshwara, D. B., "Endochronic Theory of Plasticity and Failure of Concrete," *Journal of Engineering Mechanics Division*, ASCE, Vol. 102, No. EM4, 1976, pp. 701- 722.
2.2 Bazant, Z. P., and Shieh, C. L., "Hysteretic Fracturing Endochronic Theory for Concrete," *Journal of Engineering Mechanics Division*, ASCE, Vol. 106, No. EM5, 1980, pp. 929-950.
2.3 Brandes, K., Limberger, E., and Herter, J., "Material Response to High Rate of Loading," *International Workshop on Blast-Resistant Structures - Proceedings*, Beijing, China, Oct. 14-16, 1992, pp. 1-12.

200   Shock and Impact on Structures

2.4   Chen, W. F., *Plasticity in Reinforced Concrete*, McGraw-Hill, New York, 1982, 473 pp.
2.5   Chen, W. F., and Han, D. J., *Plasticity for Structural Engineers*, Springer-Verlag, New York, 1988, 606 pp..
2.6   ACI Committee 446 (Fracture Mechanics), "State-of-Art Report - Fracture Mechanics of Concrete: Concepts, Models and Determination of Material Properties," *Fracture Mechanics of Concrete Structures*, Z. P. Bazant, (ed.), Elsevier Applied Science, London, UK, 1992, pp. 1-140.
2.7   Technical Committee on Fracture Mechanics of Concrete - Japan Concrete Institute, "Committee Report - Fracture Mechanics of Concrete Structures," *JCI Colloquium on Fracture Mechanics of Concrete Structures - Proceedings*, Mar. 30, 1990, pp. I1-I176 (in Japanese).
2.8   RILEM, 1985-TC 50-FMC, Fracture Mechanics of Concrete, "Determination of the Fracture Energy of Mortar and Concrete by Means of Three-Point Bend Tests on Notched Beams," RILEM Recommendation, *Materials and Structures*, Vol. 18, No. 106, July-Aug. 1985, pp. 285-296.
2.9   *Reports of the Second Japan-US Seminar on Finite Element Analysis of Reinforced Concrete*, Fracture Mechanics (Chapter 2), June 3-6, 1991, pp. 53-85.
2.10 Isenberg, J., Bazant, Z. P., Mindess, S., Suaris, W., and Reinhardt, H. W., "Dynamic Fracture," *Fracture Mechanics of Concrete Structures*, Z. P. Bazant, (ed.), Elsevier Applied Science, London, UK, 1992, pp. 601-609.
2.11 Sluys, L. J., and De Borst, R., "Analysis of Impact Fracture in a Double-Notched Specimen Including Rate Effects," *Fracture Mechanics of Concrete Structures*, Z. P. Bazant, (ed.), Elsevier Applied Science, London, UK, 1992, pp. 610-615.
2.12 Chung, Y. L., "The Transient Solution of Mode-I Crack Propagating with Transonic Speed," *Fracture Mechanics of Concrete Structures*, Z. P. Bazant, (ed.), Elsevier Applied Science, London, UK, 1992, pp. 633-638.
2.13 Luong, M. P., and Liu, H., "Nonlinear Dynamic Analysis of Damage in Concrete," *Fracture Mechanics of Concrete Structures*, Z. P. Bazant, (ed.), Elsevier Applied Science, London, UK, 1992, pp. 645-650.
2.14 Zielinski, A. J., "Fracture of Concrete Under Impact Loading," *Structural Impact and Crashworthiness : Volume 2 - Conference Papers*, J. Morton, (ed.), Elsevier Applied Science Publishers, Essex, UK, 1984, pp. 654-665.
2.15 Weerheijm, J., and Reinhardt, H. W., "Concrete in Impact Tensile Tests," *Structures Under Shock and Impact*, P. S. Bulson, (ed.), Elsevier, Amsterdam, 1989, pp. 29-40.
2.16 Gorst, N. J. S., and Barr, B. I. G., "Fracture of Concrete Under Torsional Impact Load Conditions," *Fracture Mechanics of Concrete Structures*, Z. P. Bazant, (ed.), Elsevier Applied Science, London, UK, 1992, pp. 616-621.
2.17 Miyamoto, A., King, M. W., and Fujii, M., "Nonlinear Dynamic Analysis of Reinforced Concrete Slabs Under Impulsive Loads," *Journal of the American Concrete Institute*, Vol. 88, No. 4, July-Aug. 1991, pp. 411-419.
2.18 Bull, J. W., (ed.), *Precast Concrete Raft Units*, Blackie and Son Ltd, London, 1991, 193 pp.
2.19 Okada, K., (ed.), *Advanced Concrete Engineering*, Kokumin Kagakusha, 1986, 314 pp., (in Japanese).
2.20 Zukas, J. A., Nicholas, T., Swift, H. F., Greszczuk, L. B., and Curran, D. R., *Impact Dynamics*, John Wiley & Sons, New York, USA, 1982, 452 pp.
2.21 Fujii, M., Miyamoto, A., Sakai, K., and Fushita, H., "Dynamic Nonlinear Modelling of Reinforced Concrete Beams Under Impulsive Loads," *Proceedings of Japan Concrete Institute 7th. Conference*, 1985, pp. 349-352, (in Japanese).
2.22 King, M. W., Miyamoto, A., and Masui, H., "Failure Criteria and Nonlinear Dynamic Analysis of Concrete Slabs Under Impulsive Loads," *Proceedings of the Japan Concrete Institute*, Vol. 12, No. 2, 1990, pp. 859-864.

2.23 King, M. W., Miyamoto, A., and Nishimura, A., "Failure Criteria and Analysis of Failure Modes for Concrete Slabs Under Impulsive Loads," *Memoirs of the Graduate School of Science and Technology*, Kobe University, No. 9-A, Mar. 1991, pp. 1-40.

2.24 Miyamoto, A., King, M. W., and Fujii, M., "Nonlinear Dynamic Analysis of Impact Failure Modes in Concrete Structures," *Fracture Mechanics of Concrete Structures*, Z. P. Bazant, (ed.), Elsevier Applied Science, London, 1992, pp. 651-656.

2.25 Hsieh, S. S., Ting, E. C., and Chen, W. F., "A Plasticity-Fracture Model for Concrete," *International Journal of Solids & Structures*, Vol. 18, No. 3, 1982, pp. 181-197.

2.26 Ottosen, N. S., "A Failure Criterion for Concrete," *Journal of Engineering Mechanics Division*, ASCE, Vol. 103, No. EM4, 1977, pp. 527-535.

2.27 Isenberg, J., and Levine, H. S., "Finite Element Analysis of Reinforced Concrete under Shock Loading," *Seminar on Finite Element Analysis of Reinforced Concrete Structures*, Japan Concrete Institute, Vol. 1, JCI-C9E, Tokyo, May 20-24, 1985, pp. 111-130.

2.28 Han, D. J., and Chen, W. F., "Constitutive Modelling in Analysis of Concrete Structures," *Journal of Engineering Mechanics Division*, ASCE, Vol. 113, No. 4, 1987, pp. 577-593.

2.29 Kupfer, H. B., and Gerstle, K. H., "Behavior of Concrete Under Biaxial Stresses," *Journal of the Engineering Mechanics Division*, ASCE, Vol. 99, No. EM4, Aug. 1973, pp. 853-866.

2.30 Miyamoto, A., King, M. W., and Masui, H., "Non-Linear Dynamic Analysis and Evaluation of Impact Resistance of Reinforced Concrete Slabs Under Impulsive Loads," *Proceedings of the Japan Concrete Institute*, Vol. 11, No. 2, 1989, pp. 643-648.

2.31 Hannant, D. J., *Fibre Cements and Fibre Concretes*, John Wiley & Sons, Chichester, UK, 1978, 219 pp.

2.32 Isobata, O., "Two-Dimensional Elastic-Plastic Analysis of Concrete Structure by the Finite Element Method," *Transactions of the Architectural Institute of Japan*, Vol. 189, No. 11, Nov. 1971, pp. 43-50 (in Japanese).

2.33 Miyamoto, A., King, M. W., Ishibashi, T.; and Mito, M., "Nonlinear Dynamic Analysis and Evaluation of Impact Resistance for FRP Reinforced Concrete Slabs," *Proceedings of the Japan Concrete Institute*, Vol. 14, No.2, 1992, pp. 643-648, (in Japanese).

2.34 Miyamoto, A., King, M. W., and Mito, M., "Nonlinear Dynamic Analysis and Evaluation of Impact Resistance for FRP Reinforced Concrete Slabs," *Concrete Research and Technology*, Japan Concrete Institute, Vol. 4, No.1, Issue 7, Jan. 1993, pp. 20-33, (in Japanese).

2.35 Bathe, K. J., and Wilson, E. L., *Numerical Methods in Finite Element Analysis*, Prentice-Hall Inc., New Jersey, 1976, 603 pp.

2.36 Clough, R. W., and Penzien, J., *Dynamics of Structures*, McGraw-Hill Kogakusha, Tokyo, 1975, 634 pp.

2.37 Tomisawa, M., *Practical Dynamic Aseismic Design Method*, Ohm Publications, 1983, (in Japanese).

2.38 Hughes, T. J. R., *The Finite Element Method (Linear Static and Dynamic Finite Element Analysis)*, Prentice-Hall, New Jersey, 1987, 503 pp.

2.39 Nobuhara, Y., Sakurai, T., and Yoshimura N., *Program Design for Finite Element Method*, Baifuukan, 1972, 193 pp., (in Japanese).

3.1 Hand, F. R., Pecknold, D. A., and Schnobrich, W. C., "Nonlinear Layered Analysis of RC Plates and Shells," *Journal of the Structural Division*, ASCE, Vol. 99, No. ST7, July 1973, pp. 1491-1505.

3.2 Lin, C. S., and Scordelis, A. C., "Nonlinear Analysis of RC Shells of General Form," *Journal of the Structural Division*, ASCE, Vol. 101, No. ST3, March 1975, pp. 523-538.

3.3 Ueda, M., and Dobashi, Y., "Nonlinear Analysis of Reinforced Concrete Slabs," *Transcription of the Architectural Institute of Japan*, No. 283, Sept. 1979, pp. 26-36, (in Japanese).

3.4 Sekiguchi, S., "Elasto-plastic Analysis of Reinforced Concrete Slabs by Finite Element Method," *Report of the Port and Harbour Research Institute*, Vol. 19, No. 2, June 1980, pp. 170-185, (in Japanese).

3.5 Washizu, K., Miyamoto, H., Yamada, Y., Yamamoto, Y., and Kawai T., *Finite Element Handbook : Vol. 1 - Basic Principles*, Baifuukan, 1981, 443 pp. (in Japanese).

3.6 Zienkiewicz, O. C., and Cheung, Y. K., *The Finite Element Method in Structural and Continuum Mechanics*, McGraw-Hill, UK, 1967, 274 pp.

3.7 Hughes, T. J. R., *The Finite Element Method - Linear Static and Dynamic Finite Element Analysis*, Prentice-Hall Inc., New Jersey, 1987, 803 pp.

3.8 Zienkiewicz, O. C., *The Finite Element Method (3rd. Edition)*, McGraw-Hill, UK, 1977, 787 pp.

3.9 Zienkiewicz, O. C., and Cheung, Y. K., "The Finite Element Method for Analysis of Elastic Isotropic and Orthotropic Slabs," *Proceedings of the Institution of Civil Engineers*, Paper No. 6726, Vol. 28, Aug. 1964, pp.471-488.

3.10 Nath, B., *Fundamentals of Finite Elements for Engineers*, Athlone Press of the University of London, London, 1974, 255 pp.

3.11 Huang, H. C., *Static and Dynamic Analyses of Plates and Shells (Theory, Software and Applications)*, Springer-Verlag, London, UK, 1989, 194 pp.

3.12 Reddy, J. N., *Energy and Variational Methods in Applied Mechanics (With an Introduction to the Finite Element Method)*, John Wiley & Sons, New York, 1984, 545 pp.

3.13 Reddy, J. N.; Krishnamoorthy, C. S.; and Seetharamu, K. N. (eds.), *Finite Element Analysis for Engineering Design (Lecture Notes in Engineering 37)*, Springer-Verlag, Berlin, Germany, 1988, 868 pp.

3.14 Harmon, T. G., and Zhangyuan, N., "Shear Strength of Reinforced Concrete Plates and Shells Determined by Finite Element Analysis Using Layered Elements," *Journal of Structural Division*, ASCE, Vol.115, No.ST5, May 1989, pp. 1141-1157.

3.15 Zararis, P. D., "State of Stress in RC Plates Under Service Conditions," *Journal of Structural Engineering*, ASCE, Vol. 112, No. 8, Aug. 1986, pp. 1908-1927.

3.16 Frantzeskakis, C., and Theillout, J. N., "Nonlinear Finite Element Analysis of Reinforced Concrete Structures with a Particular Strategy Following the Cracking Process," *Computers & Structures*, Vol. 31, No. 3, 1989, pp. 395-412.

3.17 Miyamoto, A., and King, M. W., "Nonlinear Dynamic Analysis and Evaluation of Impact Resistance for Reinforced Concrete Beams and Slabs Under Impulsive Load," *Memoirs of the Faculty of Engineering*, Kobe University, No. 36, Nov. 1989, pp. 37-62.

3.18 Miyamoto, A., King, M. W., and Fujii, M., "Improvement of Impact Resistance and Establishment of Design Concepts of Prestressed Concrete Structures," *FIP - XIth International Congress on Prestressed Concrete*, Vol. 2, Hamburg (W. Germany), June 1990, pp. T50-T53.

3.19 Bull, J. W., (ed.), *Precast Concrete Raft Units*, Blackie and Son Ltd, London, 1991, 193 pp.

3.20 Miyamoto, A., King, M. W., and Masui, H., "Non-Linear Dynamic Analysis and Evaluation of Impact Resistance of Reinforced Concrete Slabs Under Impulsive Loads," *Proceedings of the Japan Concrete Institute*, Vol. 11, No. 2, 1989, pp.643-648.

3.21 Miyamoto, A., King, M. W., Ishibashi, T., and Mito, M., "Nonlinear Dynamic Analysis and Evaluation of Impact Resistance for FRP Reinforced Concrete Slabs," *Proceedings of the Japan Concrete Institute*, Vol. 14, No.2, 1992, pp. 643-648, (in Japanese).

3.22 Miyamoto, A., King, M. W., and Mito, M., "Nonlinear Dynamic Analysis and Evaluation of Impact Resistance for FRP Reinforced Concrete Slabs," *Concrete Research and Technology*, Japan Concrete Institute, Vol. 4, No.1, Issue 7, Jan. 1993, pp. 20-33, (in Japanese).

3.23 Miyamoto, A., King, M. W., and Fujii, M., "Nonlinear Dynamic Analysis and Evaluation of Impact Resistance for FRP Reinforced Concrete Slabs," *17th Conference on Our World in Concrete & Structures*, Vol. XI, Singapore, Aug. 25-27, 1992, pp. 95-103.

3.24 Fujii, M., Miyamoto, A., Fushita, H., and Nakatsuji, J., "Analysis of Ultimate Behaviors of RC Slabs Under Impulsive Loads," *Proceedings of the Japan Concrete Institute*, Vol. 9, No. 2, pp. 609-614, (in Japanese).

3.25 King, M. W., Miyamoto, A., and Masui, H., "Failure Criteria and Nonlinear Dynamic Analysis of Concrete Slabs Under Impulsive Loads," *Proceedings of the Japan Concrete Institute*, Vol. 12, No. 2, 1990, pp. 859-864.

3.26 King, M. W., Miyamoto, A., and Nishimura, A., "Failure Criteria and Analysis of Failure Modes for Concrete Slabs Under Impulsive Loads," *Memoirs of the Graduate School of Science and Technology*, Kobe University, No. 9-A, Mar. 1991, pp. 1-40.

3.27 Miyamoto, A., King, M. W., and Fujii, M., "Analysis of Failure Modes for Reinforced Concrete Slabs Under Impulsive Loads," *Journal of the American Concrete Institute*, Vol. 88, No. 5, Sept.-Oct. 1991, pp. 538-545.

3.28 Zukas, J. A., Nicholas, T., Swift, H. F., Greszczuk, L. B., and Curran, D. R., *Impact Dynamics*, John Wiley & Sons, New York, 1982, 452 pp.

3.29 Gonzalez-Vidosa, F., Kotsovos, M. D., and Pavlovic, M. N., "Symmetrical Punching of Reinforced Concrete Slabs : An Analytical Investigation Based on Nonlinear Finite Element Modeling," *ACI Structural Journal*, May-June 1988, pp. 241-250.

3.30 Zielinski, A. J., and Reinhardt, H. W., "Impact Stress-Strain Behaviour of Concrete in Tension," RILEM, CEB, IABSE, IASS Interassociation Symposium : *Concrete Structures Under Impact and Impulsive Loading - Proceedings*, Berlin (BAM), June 2-4, 1982, pp. 112-124.

3.31 Miyamoto, A., King, M. W., and Fujii, M., "Nonlinear Dynamic Analysis of Reinforced Concrete Slabs Under Impulsive Loads," *Journal of the American Concrete Institute*, Vol. 88, No. 4, July-Aug. 1991, pp. 411-419.

3.32 Miyamoto, A., King, M. W., and Fujii, M., "Nonlinear Dynamic Analysis of Impact Falure Modes in Concrete Structures," *Fracture Mechanics of Concrete Strucures*, Z. P. Bazant, (ed.), Elsevier Applied Science, London, 1992, pp. 651-656.

3.33 Miyamoto, A., King, M. W., and Fujii, M., "Nonlinear Dynamic Layered Finite Element Procedure for Soft Impact Analysis of Concrete Slabs," *Structures Under Shock and Impact II*, P. S. Bulson, (ed.), Computational Mechanics Publications, London (UK), 1992, pp. 297-308.

3.34 King, M. W., Miyamoto, A., and Fujii, M., "Dynamic Behavior and Analysis of Reinforced Concrete Slab Structures Under Soft Impact," *International Workshop on Blast-Resistant Structures*, Beijing (P.R.China), Oct. 14-16, 1992, pp. 31-40.

3.35 King, M. W., Miyamoto, A., and Ishibashi, T., "Modeling of Impact Load Characteristics and Its Application to Analysis of RC Slab Structures," *Proceedings of the Japan Concrete Institute*, Vol. 13, No. 2, 1991, pp. 1039-1044.

3.36 King, M. W., and Miyamoto, A., "Interfacing of Impact Load Characteristic Analysis with Dynamic Response Analysis of Concrete Slab Structures," *Memoirs of the Graduate School of Science and Technology*, Kobe University, No. 10-A, Mar. 1992, pp. 41-71.

3.37 King, M. W., Miyamoto, A., and Fujii, M., "Analytical Prediction of Impact Failure Modes in Concrete Slab Structures Due to Accidental Collisions," *Proceedings of International Symposium on Natural Disaster Reduction and Civil Engineering*, Japan Society of Civil Engineers Kansai Chapter, Sept. 1991, pp.147-158.

3.38 Eibl, J., "Design of Concrete Structures to Resist Accidental Impact," *The Structural Engineer*, Vol. 65A, No. 1, Jan. 1987, pp. 27-32.

3.39 King, M. W., Miyamoto, A., and Fujii, M., "Application of Layered Finite Element Method for Analysis of Impact Failure in Concrete Structures," *Proceedings of the International Symposium on Impact Engineering*, Vol. I, Sendai, Nov. 2-4, 1992, pp. 235-240.

3.40 Hanshin Expressway Public Corporation, *Manual for Setting of Preventive Structures*, 1972, 145 pp., (in Japanese).

3.41 Kamal, M. M., and Lin, K. H., "Collision Simulation," *Modern Automotive Structural Analysis*, Mounir M. Kamal and Joseph A. Wolf, Jr., (eds.), Van Nostrand Reinhold Company, New York, 1982, pp.316-355.

4.1 Miyamoto, A., King, M. W., and Mito, M., "Nonlinear Dynamic Analysis and Evaluation of Impact Resistance for FRP Reinforced Concrete Slabs," *Concrete Research and Technology*, Japan Concrete Institute, Vol. 4, No.1, Issue 7, Jan. 1993, pp. 20-33, (in Japanese).

4.2 Miyamoto, A., and King, M. W., "Nonlinear Dynamic Analysis and Evaluation of Impact Resistance for Reinforced Concrete Beams and Slabs Under Impulsive Load," *Memoirs of the Faculty of Engineering*, Kobe University, No. 36, Nov. 1989, pp. 37-62.

4.3   Miyamoto, A., King, M. W., and Masui, H., "Non-Linear Dynamic Analysis and Evaluation of Impact Resistance of Reinforced Concrete Slabs under Impulsive Loads," *Proceedings of the Japan Concrete Institute*, Vol. 11, No. 2, 1989, pp. 643-648.

4.4   Koyanagi, W., Rokugo, T., and Horiguchi, H., "Failure Condition of Steel Fiber Reinforced Concrete Beam Element under Repeated Impact Loading," *Transactions of the Japan Cement Association*, Vol. 38, 1984, pp. 381-384, (in Japanese).

4.5 Kamil, H., Krutzik, N., Kost, G., and Sharpe, R., "An Overview of Major Aspects of the Aircraft Impact Problem," *Nuclear Engineering and Design*, Vol. 46, 1978, pp. 109-121.

Chapter 6

Impact problems in aeroengines

C. Ruiz, D. Hughes

University Technology Centre for Solid Mechanics,
Department of Engineering Science, Oxford University,
Parks Road, Oxford OX1 3PJ, UK

ABSTRACT

Modern aeroengines are designed to withstand a variety of static and dynamic loads. Impact considerations are often of paramount importance. Both hard impact, as in the case of the need to contain the fragments of a blade and soft impact which occurs when a bird is ingested, are described here.

INTRODUCTION

Modern gas turbines used nowadays in the majority of aircraft are doubtless amongst the most advanced machines. A large modern aeroengine may have typically a thrust of 400,000 N, a power exceeding 50 MW on take-off, sufficient to provide enough power to supply a town with a population of 100,000 inhabitants and a weight of the order of 20 tonnes. Even more important, this performance is accompanied by an excellent strength and reliability. The development of aeroengines as illustrated in Fig.1 is such as to demand constant improvements in spite of the already remarkable achievements of the past decades.

A typical high by-pass ratio commercial engine consists of a compressor, a combustion chamber and a turbine. Turbine and compressor are connected by a shaft or a number of concentric shafts. The first stage of the compressor is a rotor with a large number of blades which rotate within a casing at peripheral tip speeds of around 600 m/s. Some of the air sucked in by the engine is used in the turbine from which it escapes together with the combustion gases at a very high velocity and provides in this manner some of the thrust. The rest is by-passed and directed by a row of stationary guide vanes leaving the engine at a lower velocity than the exhaust gas and providing the rest of the thrust.

Given the mass and high kinetic energy of the fan blades,

Fig. 1 - Development of the aeroengine
(From A history of Technology, T.L. Williams, ed.,
Vol. VII, OUP, 1978)

Year	Engine	Power (bhp)	Weight/Power (kg/bhp)
1880	Otto engine	<20	200
1890	Daimler, Maybach	20	30-4.0
1910	Antoinette	50	2.0
	Gnome	50	1.5
1920 to 1930	Liberty (vi2) Hispano (v8) Eagle	} 400	1.0
1940 to 1960	Merlin Wright Cyclone BMW, Junkers	} <2,000	0.5
1920/30	Gas Turbine study	100	2.7
		Thrust(lb)	Weight/Thrust
1940	Whittle	1,700	≈1.0
	Junkers	2,000	
1970	RB211	38,000	0.3
1990	R-R Trent and derivatives	70.000	0.2

Fig. 2   - typical turbofan.

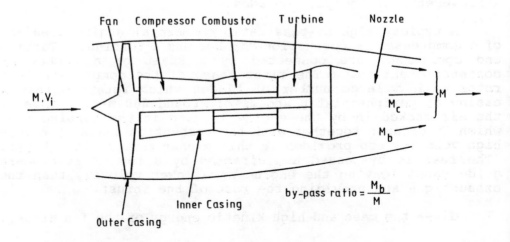

$$\text{by-pass ratio} = \frac{M_b}{M}$$

it is obvious that the rupture of the junction between blade
and rotor could have catastrophic consequences.   For this
reason the casing must be designed to contain a blade should
it break away from its root.   The problem is less severe in
the rest of the engine due to the smaller size of the blades
but it is similar in nature.   Essentially all casings must
withstand the impact of a hard missile at a velocity ranging
between 100 and 600 m/s.

       When the aircraft is flying low, there is the possibility
of ingesting with the air small objects such as pebbles, hail-
stones or sand which may cause some localised damage.   The
ingestion of a bird can be far more serious.   A sparrow can
cause some plastic deformation of the leading edge of a blade
impairing its aero-dynamic efficiency.   A goose can result in
the complete rupture of the blade root or tearing of the blade
producing the detachment of a high energy missile.

       There are therefore two problems that must be solved, the
hard impact of metal against metal and the soft impact of a
bird, usually modelled as a block of gelatine, against the
blade which may be metallic or made of a fibre-reinforced
composite.

HARD IMPACT

In the impact between the blade and the shield the damage is
always caused by the blade root.   The leading edge trails
along the shield as the blade separated from the root turns
around the first point of impact and hits the shield at a
velocity in the region of 200 m/s.   Since the shield material
is titanium or aluminium alloy, there is little risk of
brittle fracture.   The process is one of plastic deformation
which may lead to tearing.   The energy dissipation mechanisms
consist of elastic vibrations leading to damping at the
supports, plastic deformation in membrane stretching, bending
and shear, plugging and plastic deformation of the missile.
The mechanism is illustrated in Fig. 3 in which the various
terms involved in energy dissipation are listed [1,2].
Elastic vibration is governed by the relationship between
yield point and structural stiffness.   Plastic deformation
includes stretching and bending terms and leads to petalling.
Both processes take place at different velocities:
       1 - Elastic deformation with stress waves at a speed of
around 5,000 m/s, the deformation extends over a 300 - 500 mm
diameter within 30 - 50 μs.
       2 - Plastic deformation propagates at about 200 m/s.   To
cover the same field the time elapsed is of the order of 10ms.
       3 - Plugging.   For a projectile velocity of 200 m/s and
a shielding thickness of 5 mm, the time taken to perforate the
shield is around 50 μs, ie. of the same order as the time
taken for the elastic deformation of the shield.   This third

process is illustrated in Fig. 4.   It consists of an initial
compression ahead of the missile followed by the setting up of
narrow shear bands in which the material is subjected to very
high strain rate and reaches an elevated temperature.   Known
as adiabatic shear, the energy associated with the process is
extremely low and in the design of the shield it is essential
to ensure that failure should it occur is governed by overall
plastic deformation.

Fig. 3   - Competing energy dissipation mechanisms.

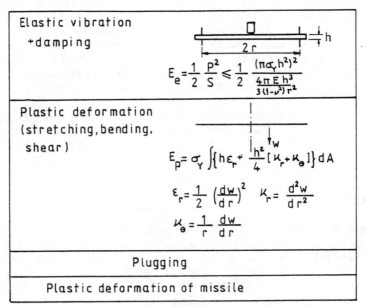

A series of tests were conducted in Oxford to ascertain
the efficiency of the shield in terms of energy per unit area.
Fig. 5 shows the results for mild steel plates.   Below a
thickness of 5 mm, the perforation energy is approximately
proportional to the square of the thickness and failure occurs
as a result of large plastic deformations.   Above 5 mm,
failure is by plugging (adiabatic shear) and the efficiency
of the shield drops significantly.   Fig. 6 illustrates a
similar result for two series of stainless steel targets in
which the transition from plastic deformation at low
velocities to plugging at high velocities and increasing
target thickness manifests itself by the presence of a "kink"
in the curves.   Also plotted in Figs. 5 and 6   are the
results obtained for shields consisting of several layers of
variable thickness.   Referring to Fig. 5, the two or three
layers with a thickness of 1.18 mm separated or clamped
together behave less well than the equivalent thickness in a

Fig. 4 - Effect of impact velocity on failure mode. At high
velocity, plugging occurs, while petalling takes
place at low velocity. Mixed mode illustrated shows
compression Zone and shear band.

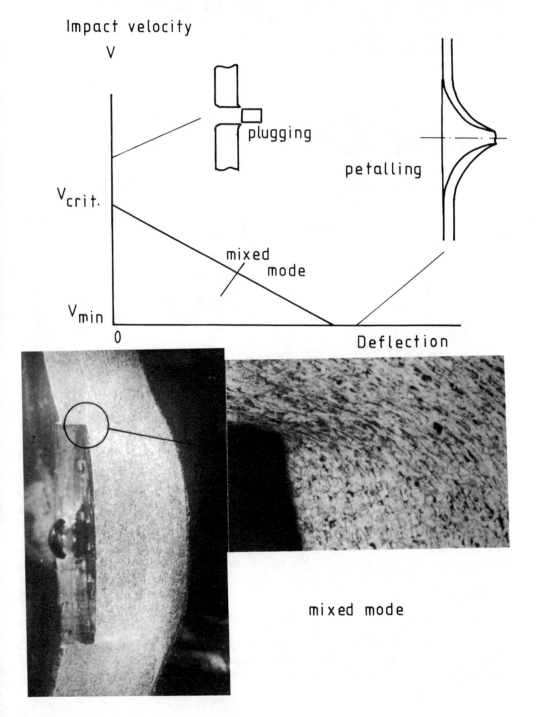

mixed mode

single layer but three layers of 1.96 mm or two of 3.00 mm show a distinctive improvement. Likewise in Fig. 6 the thick multilayers of stainless steel are better than the single layers of equal total thickness while the thinner multilayers are less effective. It should also be noted that the typical kink of the single layers has disappeared and that the multilayers always fail as a result of combined plastic deformation and plugging.

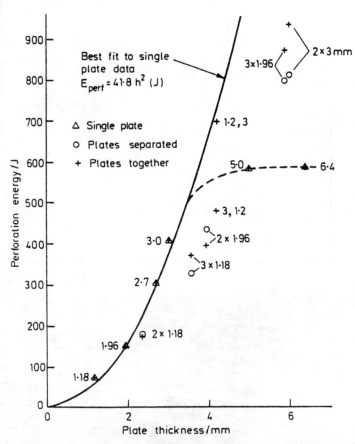

Fig. 5 – Comparison between single layer shields and multilayer. Single layer curve levels off at around 600 J between 3.0 and 5.0 mm plate. (Broken line).

It is clear that the global behaviour is the most important factor and from the experimental observations [3] it follows that whether failure occurs as a result of plastic deformation or plugging depends almost entirely on the initial response of the shield. The two main conclusions from the tests

conducted in Oxford are that:
    1 - the deceleration force varies as a function of the structural stiffness of the shield,
    2 - perforation occurs when the average shear stress reaches a critical value.

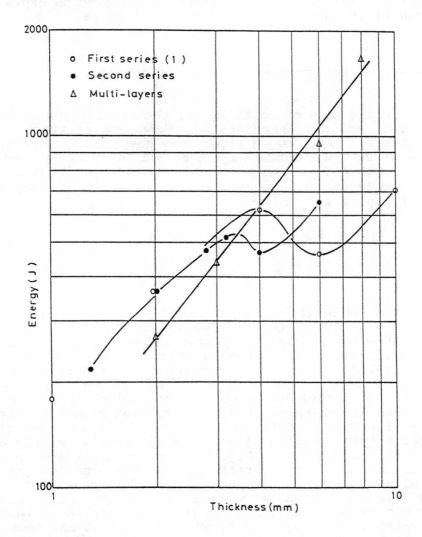

Fig. 6 - Single and multilayer stainless steel shields.

    For a projectile of diameter d at a velocity V and a target of thickness h and stiffness k, the following equation applies

$$\frac{E}{E_o}=\left(\frac{d}{d_o}\right)^n\left(\frac{k_o}{k}\right)^m$$

where $n = 1$ and $m = 1/3$ [4].   The subindex o indicates a reference shield.   The stiffness k can also be expressed in terms of the natural vibration frequency as,

$$k=\varpi^2\rho h$$

where ϖ is the frequency and ρ the density.

In the case of multilayers, the first layer is backed by the rest and has therefore an equivalent stiffness equal to the total.   The second layer only has $(n - 1)$ backing layers if n is the total number and the last layer is unbacked.   The total combined thickness is nk_1 where k_1 is the stiffness of a single layer. The absorbed energy in a shield consisting of equal layers is given by,

$$E=E_o\left(\frac{n}{n_o}\right)^{1.23}$$

The use of the two equations is illustrated in Figs. 7(a) and 8 which refer to a number of tests with targets in the form of beams or circular plates and projectiles of different masses.   Fig. 7(a) for mild steel, shows that the raw data illustrated by the open triangles and diamonds collapse onto a narrow band within + or - 20% of the average values when scaled in accordance with the preceding equations.   The collapse onto a narrow band becomes even more apparent in the case of aluminium alloy in Fig. 8 [4].   In the case of multi-layer shields scaling also brings all results within a narrow band as illustrated in Fig. 7(b) for mild steel shields.   In order to promote overall plastic deformation and avoid localised plugging through adiabatic shear, it is also possible to introduce a hard front layer of a structural ceramic such as alumina.   In this way the projectile is defeated by deforming or buckling before it has been able to penetrate into the softer, more ductile backing plates that constitute the bulk of the shield.   There are however practical problems that prevent the widespread use of a hard front layer.   An alternative is to have a very soft front layer of aluminium whose purpose is to form a cushion around the sharp corners of the projectile and thus spread the load over a wider area.   Multilayer shields consisting of aluminium and stainless steel have been shown to be remarkably

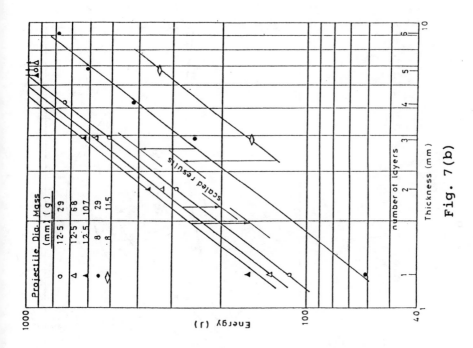

Fig. 7(b)

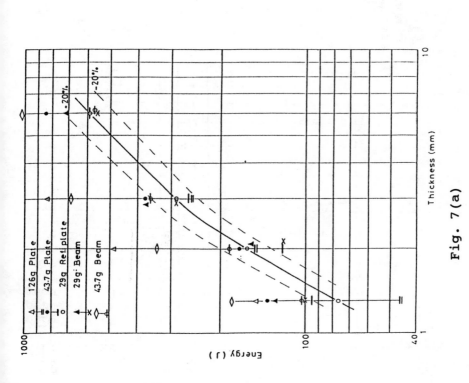

Fig. 7(a)

Comparison between reference test results and scaled
results: mild steel

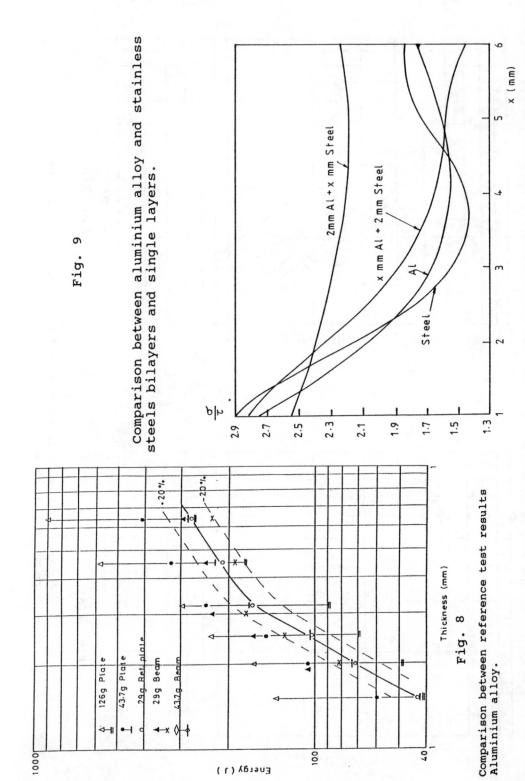

Fig. 9

Comparison between aluminium alloy and stainless
steels bilayers and single layers.

Fig. 8

Comparison between reference test results
Aluminium alloy.

effective in reducing the magnitude of the shear stress as shown in Fig. 9 [5]. The ratio between the maximum tensile stress and the shear stress when the peak force is reached indicates the tendency to deform plastically rather than to fail through plugging. The higher the ratio, the more efficient the shield for a given thickness. In Fig. 9 a shield consisting of a front layer of aluminium 2 mm thick backed by stainless steel is the most efficient over a wide range of thicknesses. The plain stainless steel is the least efficient over the thickness range between 2 and 4.5 mm.

In practice metal shields are only used nowadays for small diameter casings. The large diameter casings of high bypass ratio commercial engines normally consist of a thin metal core around which are wrapped bands of woven high strength fibre such as Kevlar [6]. The design of such shields is mainly empirical and requires a large amount of full scale testing to satisfy the civil aviation authorities.

Nothing has been said about the role of computer models or finite element analysis. Although mathematical modelling is important for the interpretation of carefully controlled experimental results its value is questionable when it comes to explaining or predicting the real complex situation.

SOFT IMPACT

One of the first collisions with a bird (birdstrikes) occurred in 1912 [7] when a seagull became entangled in the control wires of a Wright aeroplane which plunged into the sea, killing the pilot. The seriousness of birdstrikes has increased as high performance turbine have come into service. Today 60% of the birdstrikes affect the engine. Significant work has been conducted by Barber et al. [8], Bertke [9] and Boehman [10] on the effect of birdstrike on the first stage fan blades. Supplementary to this basic analytical work has also been undertaken [11, 12, 13].

The tests required by the civil aviation authorities on an engine are most demanding and are classed using small birds (2 - 4 oz), medium birds (1.5 lb) and large birds (4 lb). Each test is specifically designed for a particular size of bird. In the medium bird tests, for example, in the case of a Rolls-Royce RB 211 engine 8 birds of approximately 1.5 lb in weight are fired in rapid sequence into an engine running at maximum take-off power. Not only must the engine recover, without pilot action from any resultant surging, but it must continue to run at such a level which will allow delivery of 75% thrust for 30 minutes. The objective of this test is to demonstrate that the engine can continue to propel the aircraft and that the pilot will have sufficient time to land safely.

The impact resistance of the fan blade is the most important factor in designing the engine to achieve the best performance possible and also to pass the certification requirements.   Besides the material of the blade, the leading edge profile plays an important role in determining the mode of failure which can be:
      1 - plastic deformation (cupping)
      2 - localised tear following cupping
      3 - fracture at the root.

Generally the complex nature of soft body impact tends to require extensive full scale testing programmes.   The cost of such tests prevents their widespread use at the conceptual design stage and whilst they are necessary for certification, simplified testing is preferred to generate design data.

Analysis of Bird Impact

One of the major problems arising from the use of real birds during testing is the lack of repeatability of the results generated.   Unfortunately, although providing realism, the actual impact loads of real birds vary from test to test. The differences are caused by a number of factors, from an inability to control the orientation of the impact on differences of density, homogeneity, isotropy and symmetry; basically no two birds are exactly the same.   Added to this is the problem of scaling the loads, starlings are not exact replicas of gulls and also finding a small scale model of a starling is difficult [14].   Finally there are hygiene and sanitation problems involved in testing real birds.

Therefore, it is necessary to develop a substitute bird material which has similar impact properties to real birds.   The substitute material must be relatively inexpensive and readily available.   Extensive testing was undertaken by Wilbeck and Rand [15] in order to characterise a substitute bird model.   In the tests birds ranging in mass from 60 to 600 g and larger birds ranging from 1 to 4 kg, were launched by means of a gas gun at velocities between 50 to 300 m/s.   The birds were fired, rear end first, at flat, rigid plates with pressure transducers embedded flush with the surface.   Fig. 10 shows a typical pressure trace, in which there is an initial peak followed by approximately equal pressure.   The actual duration of the impact was equal to the time taken for the bird to travel through its length, expressed analytically as : -

$$T = \frac{L}{u_o}$$

where, T, is the duration of the impact, L is the length of

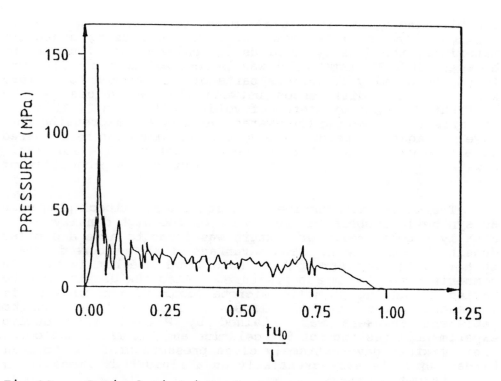

Fig.10 - Typical Bird impact pressure trace at 200 m/s velocity.

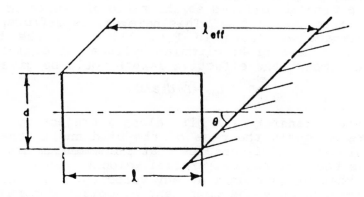

Fig.11 - Oblique impact: effective bird length.

the bird and u_o is the initial impact velocity.   In order to replicate the results from the bird tests using synthetic materials, the density of birds ie. chickens was found to be between 900 - 950 kgm^{-3} which was in contrast to Tudor [16] who found the density of various parts of the chicken to be 1060 kgm^{-3} .   The differences between the two values can be attributed to the existence of voids in the bird.   This has the effect of lowering the average density and decreasing the wave propagation speed.   The density of birds can be compared to a mixture of 85 - 90% volume of water to 10 - 15% air. However, it is extremely difficult to produce a projectile of such a mixture.

Therefore, alternatives were studied by Wilbeck [15] such as synthetic rubbers but most were of too high density. Finally a gelatine/water mixture was found to have a similar density to bird flesh.   The advantage of the mixture is that it has sufficient strength to hold its shape but is weak in comparison to the impact pressure.   Also the density can be varied by adding phenolic microballoons to the mixture to represent voids in real birds.   The ideal mixture to represent chickens was obtained by Wilbeck [15] during experimental testing of 85% gelatine and 15% micro balloons. This mixture gave extremely close pressure profiles to real birds, which behave essentially as a fluid body impacting a rigid target.

In the previous equation, T the duration of impact or "squash up time" is defined as the ratio of length of the bird over the impact velocity.   This however, is different for an oblique impact as shown in Fig. 11.   If the bird was considered to be a right circular cylinder of diameter d, in an oblique impact the effective length would be given

$$L_{eff}=L+d\tan\theta$$

The momentum transfer of a bird along a trajectory is simply mu_o, where m equals the mass of the bird and u_o the initial impact velocity.   This assumes that the momentum after impact is zero as the bird has only radial velocity since there is no evidence that birds bounce at any velocity.   In the case of oblique impacts only the component of momentum normal to the impact surface is transferred to the target during impact, therefore, the momentum transfer is given by

$$I=mu\sin\theta$$

θ is the angle between trajectory and surface of the target. The average impact force is given by

$$F_{avg}=\mu^2 L_{eff}\sin\theta$$

Using these simple relationships as non-dimensionalizing

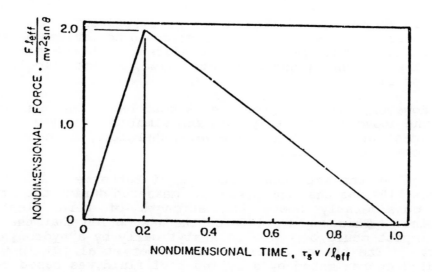

Fig.12 - Generalized bird impact force - time profile.

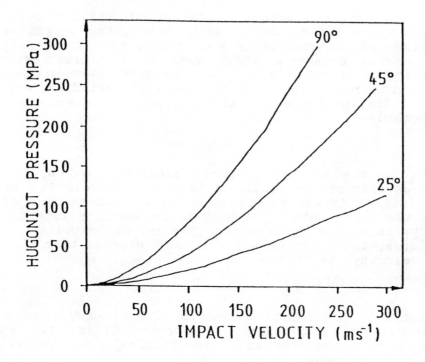

Fig.13 - Shock pressure versus impact velocity (gelatine 10% porosity)

factors, Barber et al [8] showed by experimental testing a generalised force - time bird impact could be derived as shown in Fig. 12.   This profile is consistent with the test data and takes account of the bird mass, size, impact velocity and angle of impact.

However, although providing a good basis for analytical work the above relationships are too simple to fully describe the impact process. It is necessary, therefore, to consider a more   complex analysis.

One of the important findings of Barber et al [8] and Wilbeck [15] was that the pressures measured during the impact were   considerably   above   the   strength   of   bird   flesh   or gelatine.   This would reinforce the assumption that the soft body impact could be modelled analytically by a hydrodynamic theory.   The theory developed by Barber et al [8] in which the bird approximated by a cylinder of fluid was based on previous studies of water drops and water jets impacting plates by Bowder and Brunton [17], Heyman [18] and Glen [19]

At the initial instant of impact of a fluid cylinder impinging normally against a rigid target, the particles at the surface of the projectile are brought instantaneously to rest.   This   causes   a   shock   wave   to   be   formed   at   the target/projectile interface.   As the shock wave propagates into the projectile, it brings the material behind the wave to rest.   The   pressure   at   the   surface   is   given   by   the relationship:-

$$P_H = \rho_o u_s u_o$$

where P_H is the Hugoniot (shock) pressure, and ρ_o is the initial density of the fluid, u_s is the shock velocity and u_o is the initial impact velocity.   For an impact on a rigid target, the shock velocity (which increases with impact velocity) is always greater than the impact velocity.   For most fluids, the shock velocity is much greater than the impact velocity, so that:-

$$P_H >> \rho_o u_o^2$$

For relatively low velocity impacts with $c_o >> u_o$ (where c_o is the velocity of sound in the unshocked fluid) the shock pressure can be approximated by the classical water hammer, or low velocity Hugoniot pressure [20].

The shock pressure depends only on the component of the projectile velocity normal to the target and it is therefore, easy to modify the equations to include a target angle θ. The effect of this angle on the Hugoniot pressure was studied using experimental techniques by Wilbeck [21] and Fig. 13

shows the pressure plotted against velocity for different angles of incidence.

The duration of the shock pressure at the target face depends on the shock release rate. Initially the high shock pressure exists all along the impact surface. At the boundaries of the impact surface, the material is bounded on one side by a high pressure and on the other by a free surface. This results in an outward acceleration of the material, thereby developing a pressure release wave which propagates inwards towards the centre of the impact. The release wave causes a dramatic decrease in the pressure at the target face. As a result the duration of the high shock pressure is greatest at the centre of impact and can be calculated using:-

$$t_r = \frac{r}{c_s}$$

where t_r is the time for the release wave to reach the centre, r the initial radius of the cylinder and c_s is the speed of sound in the shocked material.

After the release pressure is complete, the impact process attains a steady state condition. During this steady state, the pressure on the target surface at the centre of impact is the stagnation pressure P_{ss} given by Bernoulli's relationship:

$$\int_{P_o}^{(P_{ss}+P_o)} \frac{dP}{\rho} = \frac{u_o^2}{2}$$

P_o and u_o are the pressure and velocity of the uniform flow field some distance away from the impact surface and are approximated to the atmospheric pressure and initial impact velocity. This equation can be solved to give:-

$$P_{ss} = K\rho_o u_o^2$$

For an incompressible fluid, $K = 0.5$. However, for most materials the density tends to increase with the applied pressure, so that K will have values which approach 1.0.

In the case of oblique impact, the problem is considerably more complex. Fig. 14 shows steady state flow of an oblique impact of a cylinder on a rigid plate. From momentum considerations it can be seen that most of the fluid will flow downstream on the obtuse side of the impact. The stagnation point shifts upstream to the acute side of the centre of the impact. As long as the stagnation point exists, the full stagnation pressure is given by the preceding equation.

Therefore, the maximum pressure generated during steady state flow will be independent of angle of impact.   However, it must be noted that the pressure distribution over the surface will be greatly dependent on impact angle.

The distribution of pressure in an oblique cylindrical impact is difficult to analyze as it is a dynamic three dimensional fluid flow problem.

Leading Edge Impacts

In an investigation by Bertke [22] both hard and soft projectiles were fired at a range of different targets.   The target materials were stainless steel 403 and titanium alloy Ti-8Al-1Mo-1V.   The geometries of the titanium targets are shown in Fig. 15.   Considering just the artificial bird impacts, a series of 85 g gelatine (15% microballoon) projectiles were used.   The thick stainless steel specimens received severe plastic deformation at 240 m/s.   The thinner specimens received severe bending at 170m/s.   In both cases the specimens did not fail due to tearing and therefore, the critical velocity would be above the velocities on test.   The titanium targets were impacted at higher velocities than the stainless steel specimens and received no damage at 300 m/s and only a moderate amount of plastic deformation at 370 m/s.

Bertke and Barber [23] reported a more extensive experimental investigation of the effect of different impact parameters on the soft body impact of titanium specimens.   In their study, three distinct areas were investigated to obtain an indication in terms of the resultant damage, caused by each parameter.   Firstly the effect of projectile size and density on the impact damage was considered.   Secondly the effect of impact velocity and angle on the resultant damage was considered.   Finally the effect of specimen shape, size, boundary conditions, overall thickness and leading edge thickness was studied.

From the Bertke and Barber work, several important factors must be taken into account when designing an experiment.   The most important is the impact velocity.   The incidence angle is also important since it is the relative velocity normal to the blade which causes the damage.   The projectile density and specimen clamping conditions have little effect on the damage and should remain constant throughout the test.   The projectile size and the specimen size are possibly linked in that the specimen must be sufficiently long to avoid any boundary effect but small enough to reduce the structural bending.   The projectile size must be adjusted accordingly to produce the required results. Finally the leading edge thickness does have a significant effect on the local damage and therefore, must be taken into account in any test programme.

Fig. 16 shows a sequence from a dynamic finite element analysis using ABAQUS. The typical cupping deformation under the soft projectile is clearly illustrated and the predicted shape agrees will with the photograph of Fig. 17. As the velocity increases, the mode of failure suddenly changes to localised tear, illustrated in Fig. 18. The main limitation of most finite element codes is the lack of a well validated failure criterion. In its absence, the mathematical model only predicts increasing deformation and does not capture the critical situation that arises when the mode of failure suddenly changes.

CONCLUSIONS

In spite of the modern developments in computational techniques, the problems presented by impact loads in aeroengines still require a substantial amount of experimental work. Certification tests on full-scale components which attempt to model as accurately as possible the actual operational conditions will always be required for certification. The current tendency is to design simple tests intermediate between the conventional tests used for the measurement of mechanical properties of materials and the full-scale structural tests. These simple tests provide data at an early stage of the design process and can be used to verify mathematical models which are then used to support empirical rules. Both problems involving hard and soft impact although of prime importance to the aeroengine industry, are relevant to the design of any fast rotating machinery or to airframes. However, because of the highly specialised industrial interests that support this research, the results receive less publicity than they deserve remaining within a small circle of researchers.

Computational techniques and in particular finite element analysis have a place in the study of failures. Amongst the most widely used codes, explicit dynamic finite elements programmes such as DYNA are widely used. Although they demand considerable effort and computational capacity, they compare favourably in cost with experimental techniques.

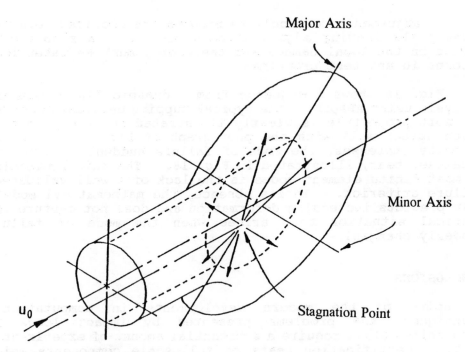

Fig.14 - Oblique impact on a rigid plane.

NOMINAL THICKNESS SPECIMEN

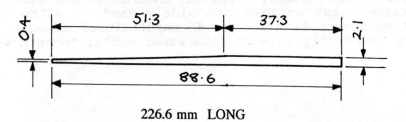

457.2 mm LONG

HALF - SCALE SPECIMEN

226.6 mm LONG

Fig.15 - Cross sectional geometry of titanium test specimens
[22]

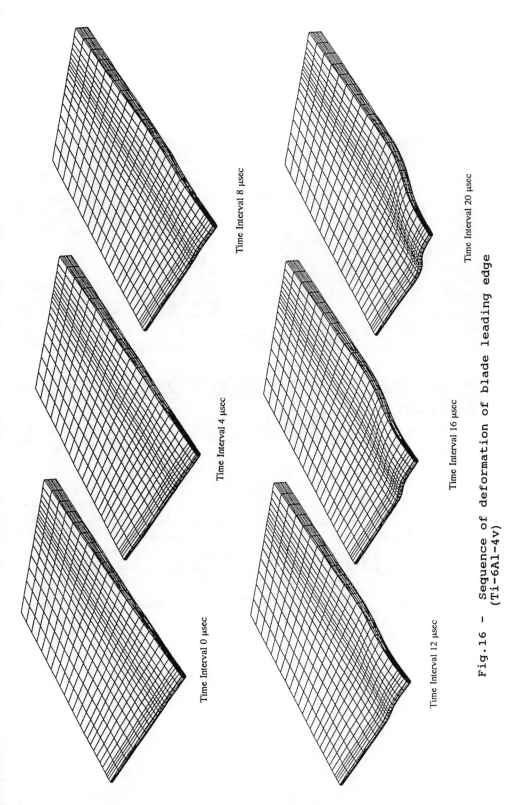

Time Interval 0 μsec

Time Interval 4 μsec

Time Interval 8 μsec

Time Interval 12 μsec

Time Interval 16 μsec

Time Interval 20 μsec

Fig.16 - Sequence of deformation of blade leading edge
(Ti-6Al-4v)

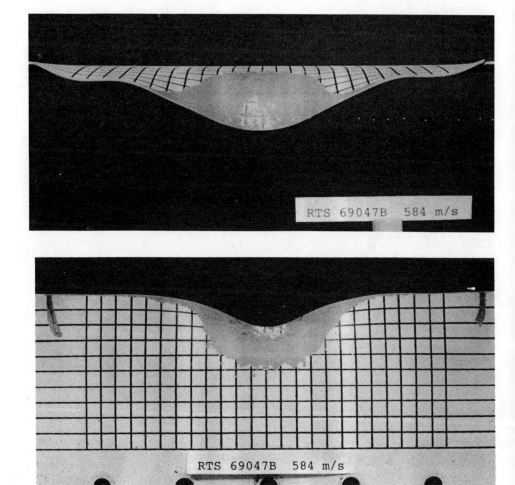

Fig.17 -   Typical deformed shape of blade leading edge.
           Cupping mode.

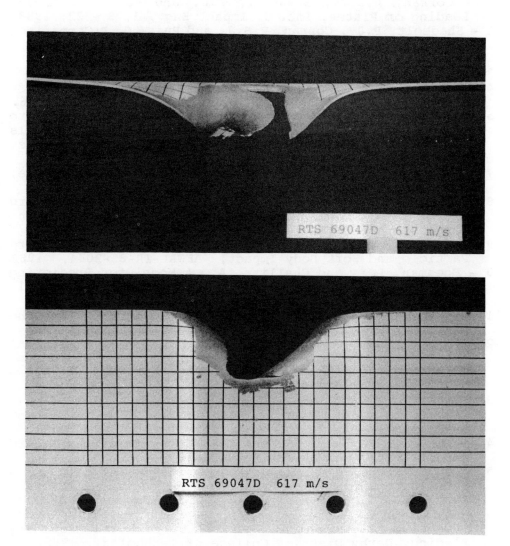

Fig.18 - Tearing failure of profile of Fig. 16 at higher impact velocity.

REFERENCES

1 - Corran, R.S.J., Shadbolt, P.J. and Ruiz, C. Impact Loading of Plates, Int. J. Impact Eng., 1, 3 - 22, 1983.
2 - Shadbolt, P.J., Corran, R.S.J. and Ruiz, C. A Comparison of Plate Perforation Models, Int. J. Impact Eng., 1, 23 - 49, 1983.
3 - Hughes, D., D. Phil. Thesis, Oxford Univ., 1992.
4 - Ruiz, C. Scaling in Penetration and Perforation Mechanics, Size Effects in Fracture, I. Mech. E., 1986.
5 - Xia, Y. and Ruiz, C. Response of Layered Plates to Projectile Impact, Proc. Int. Conf. Mech. Prop. Materials at High Rates of Strain, Oxford 1989, Inst. of Physics.
6 - Rolls-Royce private communication.
7 - Martindale, I.G. Birdstrike Familiarisation Course, Private, Communication 1989.
8 - Barber, J.P., Taylor, H.R. and Wilbeck, J.S. Bird Impact Forces and Pressures on Rigid and compliant Targets, AFFDL-TR-77-60,1978
9 - Bertke, R.S. Local Leading Edge Damage from Hard Particle and Soft Body Impact, AFWAL-TR-82-2044, 1982
10 - Boehman, L.I. and Challita, A. A Model for Predicting Bird and Ice Impact Loads on Structures, AFWL-TR-21-2046, 1982
11 - Niering, E. Simulation of Bird Strikes on Turbine Engines, DYNA3D User Group Conference, London, 1988
12 - Storace, A.F., Nimmer, R.P. and Ravenhall, R. Analytical and Experimental Investigation of Bird Impact on Fan and Compressor Blading, Journal of Aircraft, 21, 7, 520 - 527, 1984.
13 - Lawson, M. and Tuley, R. Supercomputer Simulation of a Birdstrike on a Turbofan Aero Engine, Finite Element news, 1987
14 - Challita, A. and Barber, J.P. The Scaling of Bird Impact Loads, AFFDL-TR-79-3042, 1979.
15 - Wilbeck, J.S. and Rand, J.L. The Development of a Substitute Bird Model, Journal of Engineering for Power, 103, 725 - 730, 1981
16 - Tudor, A.J. Bird Ingestion Research at Rolls-Royce, Symposium on Mechanical Reliability of Turbo-Machinery Blading, Derby District College of Technology, 1968.
17 - Bowden, F.P. and Brunton, J.H. The Deformation of Solids by Liquid Impact at Supersonic Speeds, Proc. Royal Society of London 263(a), 433 - 450, 1961
18 - Heyman, F.J. High Speed Impact between a Liquid Drop and Solid Surface, Journal of Applied Physics, 40, 13, 5112 - 5122, 1969
19 - Glen, L.A. On the Dynamics of Hypervelocity Liquid Jet Impacts on a Flat Rigid Surface, Journal of Applied Mathematics and Physics, 25, 383 - 398, 1979.
20 - Cassenti, B.N. Hugoniot Pressure Loading in Soft Body Impacts, AIAA Paper 79, 0782, 241 - 248, 1979

21 - Wilbeck, J.S.   Impact Behaviour of Low Strength Projectiles, AFML-TR-77-134, 1978.
22 - Bertke, R.S.   Local Leading Edge Damage from Hard Particle and Soft Body Impacts, AFWAL-TR-79-4019, 1979.
23 - Bertke, R.S. and Barber, J.S. Impact Damage on Titanium Leading Edges from Small Soft Body Objects, AFML-TR-79-4019, 1979

Chapter 7

Mixed finite elements for contact-impact of solids undergoing large strains

S. Cescotto, Y.Y. Zhu

Department M.S.M., Université de Liège, Quai Banning, 6, B-4000 Liège, Belgium

Abstract

This paper presents a mixed finite element procedure for dynamic analysis of nonlinear structures undergoing both geometrical nonlinearities and constitutive nonlinearities. The solid and unilateral contact finite elements are both based on a mixed variational formulation in which stresses, strains and displacements may be discretized separately. The interface behaviour is based on a penalty method and on the Coulomb dry friction law. For the appropriate time integration of the discretized equation of motion, the general explicit and implicit algorithms are suggested. Numerical results are presented for dynamic analysis such as impact of a cylinder on a rigid wall and high speed rolling.

1. Introduction

Metal-forming processes such as extrusion, rolling, upsetting, sheet forming, generally subject the workpiece to strong nonlinearities:

- large strains, large displacements and large rotations;
- material nonlinearities;
- unilateral contact boundary conditions with friction.

Another very important characteristic in some forming processes (for example: high speed rolling, impact, dynamic extrusion) is the existence of inertia forces[1-6]. With the development of finite element methods, more accurate determination of the field variables governing metal flow in forming processes which involve complex boundaries, complex frictional contact conditions, and realistic material constitutive models, becomes feasible. However, it is not very easy to find many papers published in which the solid and contact finite elements are both based on a mixed variational formulation.

Chaudhary and Bathe[1] presented a solution method for dynamic analysis of 3D contact problems with friction. Lagrange's multiplier method and Coulomb's friction were used to solve the constrained boundary conditions. The implicit time integration of the dynamic response was performed using the Newmark scheme with parameters $\beta = 0.5$ and $\gamma = 0.5$. The energy and momentum balance criteria for the contacting bodies were satisfied accurately by this method and special interface conditions to model the impacting phenomena were not needed when a reasonably small time step was used.

Chou and Wu[2] introduced an artificial damping term to the dynamic computer code DELFT that made the approach equivalent to the Dynamic Relaxation method in order to solve static metal-forming problems. Finally, they employed an explicit time integration scheme, so that the convergence problem was avoided. A slide line and a Coulomb friction between die and workpiece were postulated. Lau, Shivpuri and Chou[3] continued to develop the above work. The formulation was further extended to incorporate practical elastic-plastic material models appropriate to rolling simulation and pertinent interface contact with friction models for high speed rolling where the inertia force had a considerable effect on deformation, without taking into account the aforementioned artificial damping term.

Lindgren and Edberg[4] compared the computational efficiency of the explicit finite element code DYNA2D and the implicit code NIKE2D in the case of simulation of rolling. The outcome of this comparison was that the explicit code was preferable. It is worth mentioning that in the explicit code, a four nodes element with one integration point was used and hourglass viscosity was added in order to prevent zero-energy deformation; while for the implicit code, a four nodes element with a 2×2 Gauss quadrature rule was used, and reduced integration (i.e. one integration point) was applied to the volumetric energy.

Kanto and Yagawa[5] presented a finite element formulation for dynamic contact problems taking large deformation into account, and used a three-nodes contact element applicable to problems with a large sliding component. A trial and error algorithm was used to obtain the converged contact state. Regarding the conditions of velocity and acceleration on the contact surfaces under dynamic loading, corrections of velocity and acceleration were proposed.

Currently, the non-linear finite element analysis program LAGAMINE is being developed in the MSM department of the University of Liège, Belgium. It has been applied successfully to the static and dynamic analysis of solids with very large strains, large displacements, large rotations, and contact interfaces[6-10]. The explicit and implicit time integration schemes have been implemented. A large number of constitutive laws have been introduced. In order to avoid spurious (hourglass) zero-energy modes and locking phenomena for the solid element and to control the average overlapping between the solid boundary and foundation, Cescotto et al. have developed solid[8] and contact[9] elements based on mixed variational principles.

The goal of this paper is to present the basic aspects of the mixed formulation both for solid and contact elements as well as to show some applications in impact and dynamic analysis of non-linear structures. The main advantages of this procedure are:

•   for the mixed solid element, it has no spurious singular mode (zero-energy mode) and locking phenomena, and it is computationally very cheap;

- for the mixed contact element, the pertinent variables (contact stresses, relative velocity as well as penetrating distance between the solid and foundation) are computed only at the integration point of the mixed contact element. Hence, equilibrium and compatibility are obtained in a weighted average sense on the contact surface. Therefore, special interface conditions to model the impact phenomena are not needed[13,15] when a reasonably small time step is used. On the other hand, the average overlapping between the solid and foundation can be controlled (see Fig. 1).

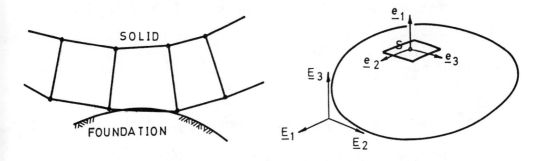

Figure 1 : Overlapping between solid and foundation

Figure 2 : Local and global reference frames

2. Mixed variational principle

2.1. Hypotheses

For the sake of simplicity, the mixed formulation will be presented first in a restrictive frame: linear elasticity and sticking contact on a known contact surface between an elastic solid and a rigid foundation. Then, all the restrictive hypotheses will be removed and the formulation will be extended to large inelastic strains and unknown contact surface with sliding and friction.

In a solid of volume V and boundary A, let A_U, A_T and A_C be the parts of A on which displacements, surface tractions and unilateral contact conditions are imposed respectively. At each point S of A_C, local co-ordinates are defined (Fig. 2), with $\underline{e}_1$ normal and $\underline{e}_2$, $\underline{e}_3$ tangent to A_C, which is assumed to be smooth. If, at point S of A_C, $\underline{\sigma}_S$ are the Cauchy stresses in the solid, (p, τ_1, τ_2) the contact stresses, $\underline{u}_S$ the displacements of the solid and $\underline{u}_F$ those of the foundation, then the following notations will be used:

$$\underline{\sigma}_S^T = <\sigma_{S11}, \sigma_{S21}, \sigma_{S31}>; \qquad \underline{\sigma}_C^T = <p, \tau_1, \tau_2>;$$

$$\underline{u}_S^T = <u_{S1}, u_{S2}, u_{S3}>; \quad \underline{u}_F^T = <u_{F1}, u_{F2}, u_{F3}>; \quad \underline{\varepsilon}_C = \underline{u}_S - \underline{u}_F,$$

where the components are evaluated in $(\underline{e}_1, \underline{e}_2, \underline{e}_3)$.

On the other hand, for points belonging to A_U, A_T or V, a global reference system $\underline{E}_1$, $\underline{E}_2$, $\underline{E}_3$ is used to express the components of stresses $\underline{\sigma}$, strains $\underline{\varepsilon}$, displacements $\underline{u}$, surface tractions $\underline{T}$, body forces $\underline{F}$, inertia force $\rho\underline{\ddot{u}}$ (ρ is the mass density), etc.

$$\underline{\sigma}^T = <\sigma_{11}, \sigma_{22}, \sigma_{33}, \sigma_{23}, \sigma_{13}, \sigma_{12}>; \quad \underline{\varepsilon}^T = <\varepsilon_{11}, \varepsilon_{22},..., 2\varepsilon_{12}>;$$

$$\underline{u}^T = <u_1, u_2, u_3>; \quad \rho\underline{\ddot{u}}^T = <\rho\ddot{u}_1, \rho\ddot{u}_2, \rho\ddot{u}_3>; \quad \underline{T}^T = <T_1, T_2, T_3>;$$

$$\partial\underline{u}^T = <\frac{\partial u_1}{\partial X_1}, \frac{\partial u_2}{\partial X_2}, ..., \frac{\partial u_1}{\partial X_2} + \frac{\partial u_2}{\partial X_1}>; \quad \underline{F}^T = <F_1, F_2, F_3>;$$

$$\underline{\sigma}_n^T = <n_j\sigma_{j1}, n_j\sigma_{j2}, n_j\sigma_{j3}>; \quad \partial\underline{\sigma}^T = <\partial_j\sigma_{j1}, \partial_j\sigma_{j2}, \partial_j\sigma_{j3}>$$

where $\underline{n} = n_i \underline{E}_i$ is the outer normal to A_T or A_U, and $\partial_j = \partial/\partial X_j$.

2.2. Sticking contact law

Using a penalty formulation, the sticking contact law is[7]:

$$\underline{\sigma}_C = \underline{K}_C \underline{\varepsilon}_C \qquad\qquad (2.1.)$$

where $\underline{K}_C$ is a diagonal matrix of penalty coefficients. The corresponding contact strain energy density W_C and complementary energy W_C^* are:

$$W_C (\underline{\varepsilon}_C) = \frac{1}{2} \underline{\varepsilon}_C^T \underline{K}_C \underline{\varepsilon}_C \qquad\qquad (2.2.)$$

$$W_C^* (\underline{\sigma}_C) = \underline{\sigma}_C^T \underline{\varepsilon}_C - W_C = \frac{1}{2} \underline{\sigma}_C^T \underline{K}_C^{-1} \underline{\sigma}_C \qquad\qquad (2.3.)$$

2.3. The functional Π

The following functional is defined:

$$\Pi = \int_{A_C} \left[\underline{\sigma}_C^T \, (\underline{u}_S - \underline{u}_F) - W_C^* \, (\underline{\sigma}_C) \right] dA_C$$

$$+ \int_V \left[W(\underline{\varepsilon}) + \underline{\sigma}^T \, (\partial \underline{u} - \underline{\varepsilon}) - \underline{F}^T \, \underline{u} - \frac{1}{2} \, \rho \, \underline{\dot{u}}^T \, \underline{\dot{u}} \right] dV \qquad (2.4.)$$

$$- \int_{A_T} \underline{T}^T \, \underline{u} \, dA_T - \int_{A_U} \underline{\sigma}_n^T \, (\underline{u} - \underline{\tilde{u}}) \, dA_U$$

in which $\underline{\tilde{u}}$ are the displacements imposed on A_u and $W(\underline{\varepsilon}) = 1/2 \, \underline{\varepsilon}^T \, \underline{C}_{el} \, \underline{\varepsilon}$ is the elastic strain energy with $\underline{C}_{el}$ the elastic moduli tensor.

In this functional, the independent fields are $\underline{\sigma}_C$, $\underline{\sigma}$, $\underline{u}$, $\underline{\varepsilon}$. If the integral on $\underline{A}_C$ is dropped, it coincides with the Hu-Washizu principle. Taking the variation of Π with respect to the independent fields gives the following natural condition:

. compatibility on A_C : $\qquad\qquad \underline{u}_S - \underline{u}_F = \underline{K}_C^{-1} \, \underline{\sigma}_C = \underline{\varepsilon}_C$

. equilibrium on A_C : $\qquad\qquad\qquad \underline{\sigma}_C + \underline{\sigma}_S = 0$

. constitutive law in V : $\qquad\qquad\quad \underline{\sigma} = \underline{C}_{el} \, \underline{\varepsilon}$

. compatibility in V : $\qquad\qquad\qquad \underline{\varepsilon} = \partial \underline{u}$

. equilibrium in V : $\qquad\qquad\qquad \partial \underline{\sigma} + \underline{F} - \rho \, \ddot{u} = 0$

. compatibility on A_U : $\qquad\qquad\quad \underline{u} = \underline{\tilde{u}}$

. equibilibrium on A_T : $\qquad\qquad\quad \underline{\sigma}_n = \underline{T}$

3. Mixed finite elements

3.1. Discretization

Starting from the functional presented above, it is possible to develop mixed finite elements for the solid and for the contact surface. Hereafter, the presentation is limited to the plane strain state but the extension to the three-dimensional case is possible.

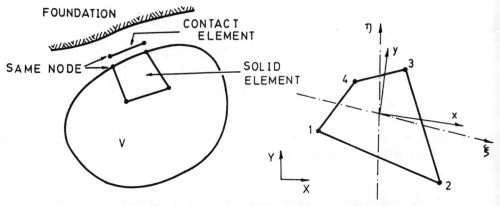

Figure 3 : Solid and contact
elements

Figure 4 : Local axes in the solid
element

The solid is discretized into 4 node quadrilaterals and the contact surface into 2 node
rectilinear elements, the nodes of which coincide with those of the underlying solid
element (Fig. 3).

3.2. Solid element

For a given element, the local axes *(x, y)* are chosen in such a way that they make equal
angles with the natural coordinates *(ξ, η)* of the element (Fig. 4). The basic assumptions
of the discretization of $\underline{u}$, $\underline{\sigma}$ and $\underline{\varepsilon}$ are the following ones. The displacements $\underline{u}$ are given
by classical bi-linear functions of *(ξ, η)*:

$$u_i = \underline{N}_S^T \, \underline{U}_i \qquad (i = x, y) \tag{3.1.}$$

with the interpolation functions:

$$\underline{N}_S^T = <N_1, N_2, N_3, N_4>, \quad N_I = \frac{1}{4} \, (1 + \xi_I \, \xi) \, (1 + \eta_I \, \eta) \tag{3.2.}$$

ξ_I, η_I are the natural co-ordinates of node I and $\underline{U}_i$ represent the nodal displacements

$$\underline{U}_i^T = <U_i^1, U_i^2, U_i^3, U_i^4> \qquad (i = x, y)$$

A more useful expression of the displacement field emerges naturally from (3.1) (3.2) by
introducing the rigid-body motion and the constant strain components into the

displacement field :[11]

$$\begin{bmatrix} u_x = a_o^x + a_x^x \, x + a_y^x \, y + a_3^x \, h \\ u_y = a_o^y + a_x^y \, x + a_y^y \, y + a_3^y \, h \end{bmatrix} \qquad (3.3.)$$

where

$$h = \frac{A}{4} \, \xi \, \eta \qquad (3.4.)$$

with A the area of the element. After some calculation, the components of the displacement gradients $\partial \underline{u}$ are found:

$$u_{i, j} = [\underline{b}_j^T + h_{, j} \, \underline{\gamma}^T] \, \underline{U}_i \qquad (i = x, y; \, j = x, y) \qquad (3.5.)$$

where

$$\underline{b}_1^T = \frac{1}{2A} \, \langle y_{24}, y_{31}, y_{42}, y_{13} \rangle, \qquad \underline{b}_2^T = \frac{1}{2A} \, \langle x_{42}, x_{13}, x_{24}, x_{31} \rangle$$

$$2A = x_{31} \, y_{42} + x_{24} \, y_{31}$$

$$x_{IJ} = x_I - x_J, \qquad y_{IJ} = y_I - y_J$$

$$\underline{\gamma} = \frac{1}{A} \, [\underline{h} - (\underline{h}^T \, \underline{x}_j) \, \underline{b}_j]$$

$$\underline{h}^T = \langle 1, -1, 1, -1 \rangle$$

x_I, y_J the co-ordinates of node I.

For the strains $\underline{\varepsilon}$, the 'optimal incompressible' option[8,11] is chosen:

$$
\begin{vmatrix} \varepsilon_x \\ \varepsilon_y \\ \gamma \end{vmatrix} = \begin{vmatrix} \bar{\varepsilon}_x \\ \bar{\varepsilon}_y \\ \bar{\gamma} \end{vmatrix} + \begin{vmatrix} h_{,x} & -h_{,y} \\ -h_{,x} & h_{,y} \\ 0 & 0 \end{vmatrix} \begin{vmatrix} \varepsilon_1 \\ \varepsilon_2 \end{vmatrix} = \underline{\bar{\varepsilon}} + \underline{h}_{,\alpha}\, \underline{\varepsilon}^* \tag{3.6.}
$$

A superposed bar designates a constant field. With this assumption, the volumetric strain is constant over the element, so the element will not lock in case of incompressibility (which is the limiting case of ideal plasticity).

For the stresses $\underline{\sigma}$, the same kind of assumption is used :[8]

$$
\begin{vmatrix} \sigma_x \\ \sigma_y \\ \tau \end{vmatrix} = \begin{vmatrix} \bar{\sigma}_x \\ \bar{\sigma}_y \\ \bar{\tau} \end{vmatrix} + \begin{vmatrix} h_{,x} & -h_{,y} \\ -h_{,x} & h_{,y} \\ 0 & 0 \end{vmatrix} \begin{vmatrix} \sigma_1 \\ \sigma_2 \end{vmatrix} = \underline{\bar{\sigma}} + \underline{h}_{,\alpha}\, \underline{\sigma}^* \tag{3.7.}
$$

The volumetric stress is constant over the element.

It is seen from (3.6)(3.7) that the constant fields $(\underline{\bar{\sigma}}, \underline{\bar{\varepsilon}})$ are uncoupled with $(\underline{\sigma}^*, \underline{\varepsilon}^*)$ which are called anti-hourglass mode.

3.3. Contact element

For a given element, local co-ordinates are chosen as indicated in Fig. 5. The displacements $\underline{u}_S$ are discretized by:

$$
\underline{u}_S = \underline{N}_C^T\, \underline{U}_C \tag{3.8.}
$$

with

$$
\underline{N}_C^T = \begin{bmatrix} \dfrac{1}{2}(1-\xi) & 0 & \dfrac{1}{2}(1+\xi) & 0 \\[2mm] 0 & \dfrac{1}{2}(1-\xi) & 0 & \dfrac{1}{2}(1+\xi) \end{bmatrix}
$$

and where $\underline{U}_C$ represent the nodal displacements of the contact element

$$
\underline{U}_C^T = [U_x^1,\ U_y^1,\ U_x^2,\ U_y^2]
$$

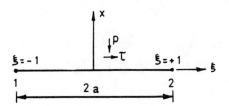

Figure 5 : Local axes in the contact element

The contact stresses are assumed to be constant over the element

$$\underline{\sigma}_C^T = \langle p, \tau \rangle \qquad (3.9.)$$

where p is the contact pressure and τ the contact shear stress. Furthermore, it is assumed that the motion $\underline{u}_F$ of the foundation is a simple translation

$$\underline{u}_F^T = \langle u_{Fx}, u_{Fy} \rangle \qquad (3.10.)$$

3.4. Variation of the discretized functional Π

From the assumptions in the solid and contact element, it is seen that the discretization parameters are :

$\underline{U}_i \ (i = x, y)$: the nodal displacements of the solid element,

$\overline{\underline{\varepsilon}}$ and $\underline{\varepsilon}^*$: appearing in the strain field $\underline{\varepsilon}$ of the solid element,

$\overline{\underline{\sigma}}$ and $\underline{\sigma}^*$: appearing in the stress field $\underline{\sigma}$ of the solid element,

$\underline{U}_C$: the nodal displacements of the contact element,

$\underline{\sigma}_C$: the (constant) contact stresses in the contact element.

Introducing these assumptions in Π (2.4) and taking its variation gives:

$$\delta\Pi = \delta\Pi_C + \delta\Pi_S = 0 \qquad (3.11.)$$

with

$$\delta\Pi_C = \delta\underline{\sigma}_C^T \, [\underline{M}_C \, \underline{U}_C - \underline{\chi}_C \, \underline{\sigma}_C - \underline{V}_F] + \delta\underline{U}_C^T \, \underline{M}_C^T \, \underline{\sigma}_C \qquad (3.12.)$$

$$\delta\Pi_S = A \, [\delta\underline{\varepsilon}^T \, (\underline{C}_{el} \, \bar{\underline{\varepsilon}} - \bar{\underline{\sigma}}) - \delta\bar{\underline{\sigma}}^T \, (\bar{\underline{\varepsilon}} - \bar{\underline{B}} \, \bar{\underline{U}}) + \delta\bar{\underline{U}}^T \, \bar{\underline{B}}^T \, \bar{\underline{\sigma}}] + \delta\bar{\underline{U}}^T \, \bar{M} \, \ddot{\underline{U}}$$
$$+ \, \delta\underline{\varepsilon}^{*T} \, [4 \, G \, \underline{H} \, \underline{\varepsilon}^* - 2 \, \underline{H} \, \underline{\sigma}^*] - \delta\underline{\sigma}^{*T} \, [2 \, \underline{H} \, \underline{\varepsilon}^* - \underline{H} \, \underline{\Gamma} \, \underline{U}] \qquad (3.13.)$$
$$+ \, \delta\underline{U}^T \, \underline{\Gamma}^T \, \underline{H} \, \underline{\sigma}^* - \delta\underline{U}^T \, \underline{F}^{ext}$$

where

$$\underline{M}_C = \begin{bmatrix} 1 & 0 & 1 & 0 \\ 0 & 1 & 0 & 1 \end{bmatrix} \qquad \underline{\chi}_C = 2a \begin{bmatrix} 1/K_p & 0 \\ 0 & 1/K_\tau \end{bmatrix} \qquad (3.14.)$$

$$\underline{V}_F = 2a \begin{bmatrix} u_{Fx} \\ u_{Fy} \end{bmatrix} \qquad \underline{H} = \begin{bmatrix} H_{xx} & -H_{xy} \\ -H_{xy} & H_{yy} \end{bmatrix}$$

$$H_{ij} = \int_A h_{,i} \, h_{,j} \, dA, \qquad (i,j = x,y) \qquad \underline{M} = \int_A \rho \, \underline{N}_S^T \, \underline{N}_S \, dA$$

$$\bar{\underline{B}} = \begin{bmatrix} \underline{b}_1^T & 0 \\ 0 & \underline{b}_2^T \\ \underline{b}_2^T & \underline{b}_1^T \end{bmatrix} \qquad \underline{\Gamma} = \begin{bmatrix} \underline{\gamma}^T & 0 \\ 0 & \underline{\gamma}^T \end{bmatrix} \qquad \underline{U}^T = [\underline{U}_x^T, \, \underline{U}_y^T]$$

G is the shear modulus, $\underline{C}_{el}$ is the plane strain elastic tensor, K_p, K_τ are the contact penalty coefficients on p and τ respectively and $\underline{F}^{ext}$ includes the body force $\underline{F}$ and surface traction $\underline{T}$.

In the preceding results, $\delta\Pi_C$ is the part of the functional which is responsible for the contact surface while $\delta\Pi_S$ comes from the discretization of the solid.

It must be noted that, since the nodes of the contact element coincide with some nodes of the solid element, $\underline{U}_C$ and $\underline{U}$ are not independent (see Fig. 3). This can be expressed by:

$$\underline{U}_C = \underline{A} \, \underline{U} \qquad (3.15.)$$

where $\underline{A}$ is an assembly operator.

From $\delta\underline{\sigma}_C$, we obtain:

$$\underline{M}_C \, \underline{U}_C - \underline{X}_C \, \underline{\sigma}_C - \underline{V}_F = 0$$

or

$$\underline{\sigma}_C = \underline{X}_C^{-1} \, [\underline{M}_C \, \underline{U}_C - \underline{V}_F] \tag{3.16.}$$

This expresses the average compatibility at contact.

From $\delta \overline{\underline{\varepsilon}}$ and $\delta \underline{\varepsilon}^*$, we obtain:

$$\overline{\underline{\sigma}} = \underline{C}_{el} \, \overline{\underline{\varepsilon}} \quad and \quad \underline{\sigma}^* = 2G \, \underline{\varepsilon}^* \tag{3.17.}$$

which expresses the elastic constitutive laws associated with the stress and strain discretization parameters.

From $\delta \overline{\underline{\sigma}}$ and $\delta \underline{\sigma}^*$, we obtain:

$$\overline{\underline{\varepsilon}} = \underline{B} \, \underline{U} \quad and \quad 2\underline{\varepsilon}^* = \Gamma \, \underline{U} \tag{3.18.}$$

which are the discretized strain-displacement relations.

Taking account of (3.16)-(3.18), $\delta\Pi$ becomes :

$$\delta\Pi = \delta\underline{U}_C^T \, \underline{F}_C + \delta\underline{U}^T \, (\underline{F}_S + \underline{M} \, \underline{\ddot{U}} - \underline{F}^{ext}) = 0 \tag{3.19.}$$

where $\underline{F}_C$ are the nodal internal forces of the contact element,

$$\underline{F}_C = \underline{M}_C^T \, \underline{X}_C^{-1} \, \underline{M}_C \, \underline{U}_C - \underline{M}_C^T \, \underline{X}_C^{-1} \, \underline{V}_F \tag{3.20.}$$

and $\underline{F}_S$ are the nodal internal forces of the solid element

$$\underline{F}_S = A \, \overline{\underline{B}}^T \, \overline{\underline{\sigma}} + \Gamma^T \, \underline{H} \, \underline{\sigma}^* \tag{3.21.}$$

Finally, using (3.15), $\delta\Pi$ becomes

$$\delta\Pi = \delta\underline{U}^T \, (\underline{F}^{int} + \underline{M} \, \underline{\ddot{U}} - \underline{F}^{ext}) = 0 \tag{3.22.}$$

with

$$\underline{F}^{int} = \underline{F}_S + \underline{A}^T \, \underline{F}_C \tag{3.23.}$$

The stiffness matrix is computed by:

$$\underline{K} \quad \frac{\partial \underline{F}^{int}}{\partial \underline{U}} - \frac{\partial \underline{F}^{ext}}{\partial \underline{U}} + \underline{M} \frac{\partial \underline{\ddot{U}}}{\partial \underline{U}} \tag{3.24.}$$

$$\underline{K}^{int} = \frac{\partial \underline{F}^{int}}{\partial \underline{U}} = A \, \underline{\bar{B}}^T \, \underline{C}_{el} \, \underline{\bar{B}} + G \, \underline{\Gamma}^T \, \underline{H} \, \underline{\Gamma} + \underline{A}^T \, \underline{M}_C^T \, \underline{\chi}_C^{-1} \, \underline{M}_C \, \underline{A} \tag{3.25.}$$

In expression (3.25), the first term corresponds to the constant field of the mixed solid element; the second term is the stabilization matrix; the third term is the stiffness matrix of the mixed contact element.

In fact, equations (3.23) and (3.25) correspond to the assembly of the contact element with the solid element. These results conclude the development of the mixed elements in the restrictive frame. In the following section, the restrictive hypotheses are removed.

It is worth noting that the stresses (3.16) and the nodal internal forces (3.20) of the mixed contact element are obtained in a weighted average sense on the contact surface. The interface stresses (3.16) depend on the penalty coefficients (see 3.14), and their values are adjusted by the equilibrium condition (3.22) if step by step methods with or without iteration (see following sections) are used. In methods where contact conditions are treated at the node level with the help of Lagrangian multipliers, it may be necessary to include special impact and release conditions.[13, 14]

However, in the present approach, this is not necessary since it is based on a penalty method and on a mixed formulation which allows a slight penetration between the solid and the foundation. This has also been indicated by other authors.[13, 15]

4. Extension of the theory

4.1. Corotational formulation

If the solid is submitted to large inelastic strains and if there is sliding with friction at contact, the formulation has to be modified.

In the solid, the local axes (x, y) are updated at each step of the calculation, in such a way that the average rotation in local axes is equal to zero during the step. With a mid point formula, this gives[8] for the step $N \to N+1$:

$$\left(\frac{\partial \Delta u_x}{\partial y}^N \quad \frac{\partial \Delta u_x}{\partial y}^{N+1} \right) - \left(\frac{\partial \Delta u_y}{\partial x}^N + \frac{\partial \Delta u_y}{\partial x}^{N+1} \right) = 0 \tag{4.1.}$$

where Δu_x, Δu_y are the incremental displacements during the step. Since they also depend on the choice of the local axes, the use of (4.1) is not straightforward. However, after some elementary algebra, the angle θ between the x local axis in $N+1$ and the X global axis is found to be given by:

$$tg\ \theta = \frac{-x_{24}^N\ X_{31}^{N+1} + x_{31}^N\ X_{24}^{N+1} + y_{42}^N\ Y_{31}^{N+1} - y_{31}^N\ Y_{42}^{N+1}}{x_{24}^N\ Y_{31}^{N+1} + x_{31}^N\ Y_{42}^{N+1} + y_{42}^N\ X_{31}^{N+1} + y_{31}^N\ X_{24}^{N+1}}$$

Then, all the equations are expressed in this local frame in the current configuration of the solid. Provided Cauchy stresses in this local frame are used, all equations remain the same as in infinitesimal formulation, except for the stress-strain relation which becomes:

$$\dot{\underline{\sigma}} = \underline{C}_T\ \dot{\underline{\varepsilon}} \tag{4.2.}$$

with $\underline{C}_T$ the inelastic compliance tensor, $\dot{\underline{\sigma}}$ and $\dot{\underline{\varepsilon}}$ the stress and strain rates in the local frame.

This corotational formulation ensures objectivity of the rate constitutive equation (4.2.) and allows the element to pass the large strain patch test.[8]

At contact also a corotational formulation is used. The local frame $(\underline{e}_1, \underline{e}_2, \underline{e}_3)$ rotates with the corresponding surface element. In this frame, the incremental contact law with sliding and friction becomes[7]

$$\dot{\underline{\sigma}}_C = \underline{K}_{CT}\ \dot{\underline{\varepsilon}}_C \tag{4.3.}$$

with

$$\dot{\underline{\varepsilon}}_C = \dot{\underline{u}}_S - \dot{\underline{u}}_F \tag{4.4.}$$

The contact constitutive tensor $\underline{K}_{CT}$ is usually non symmetric. It is based on a penalty method, on the Coulomb's dry friction law, and is developed using the elastic-plastic formalism.[7]

4.2. Mixed variational formulation

In nonlinear analysis, the functional Π is replaced by the following variational form expressed in the current deformed configuration.

$$\int_{a_C} [\underline{\sigma}_C^T \, \delta\underline{u}_S + \underline{u}_S - \underline{u}_F - \underline{\varepsilon}_C)^T \, \delta\underline{\sigma}_C] \, da_C$$
$$+ \int_v [(\partial\underline{u} - \varepsilon)^T \, \partial\underline{\sigma} + \underline{\sigma}^T \, \delta\partial\underline{u} + \rho \, \ddot{\underline{u}}^T \, \delta\underline{u} - \underline{F}^T \, \delta\underline{u}) \, dv \qquad (4.5.)$$
$$- \int_{a_T} \underline{T}^T \, \delta\underline{u} \, da_T - \int_{a_u} [\underline{\sigma}_n^T \, \delta\underline{u} + \underline{u}^T \, \delta\underline{\sigma}_n] \, da_u = 0$$

where $\underline{\varepsilon}$ and $\underline{\varepsilon}_C$ are functions of $\underline{\sigma}$ and $\underline{\sigma}_C$ obtained by integration of (4.2) and (4.3) along the equilibrium path:

$$\underline{\varepsilon} = \int_0^t \underline{C}_T^{-1} \, \underline{\dot{\sigma}} \, dt \qquad (4.6.)$$

$$\underline{\varepsilon}_C = \int_0^t \underline{K}_{CT}^{-1} \, \underline{\dot{\sigma}}_C \, dt \qquad (4.7.)$$

and similarly

$$\underline{u}_S - \underline{u}_F = \int_0^t (\dot{\underline{u}}_S - \dot{\underline{u}}_F) \, dt \qquad (4.8.)$$

In (4.5), the integrals are taken over the current configuration. The variational equation (4.5) generalizes the approach of Section 3.4 to the case of frictional contact between a rigid foundation and an inelastic solid undergoing large strains. The current contact surface a_C is determined by a simple search algorithm[10] at each step of the analysis. Finally, the variational equation (4.5) gives a discretized equilibrium equation similar to (3.22)

$$[\underline{F}^{int} + \underline{M} \, \ddot{\underline{U}} - \underline{F}^{ext}] = 0 \qquad (4.9.)$$

but which is nonlinear. Its solution can be obtained by the step by step method. Hence, the extension of the preceding elements to the case of large inelastic strains and unilateral contact with sliding and friction is obtained.

5. Equation of motion and its time integration

5.1. Equation of motion

If we consider Rayleigh-Ritz damping, the equation of motion (4.9) becomes:

$$\underline{M}\,\underline{\ddot{U}} + \underline{C}\,\underline{\dot{U}} = \underline{F}^{\,ext} - \underline{F}^{\,int} \qquad (5.1.)$$

where,

$\underline{\ddot{U}}, \underline{\dot{U}}$: vectors of nodal accelerations and velocities respectively
$\underline{M}, \underline{C}$: mass and damping matrices respectively

The internal forces include the contributions of not only the solid but also the contact finite elements, as explained in the above sections.

Equation of motion (5.1) must clearly be integrated forward in time to produce the dynamic response. Both explicit and implicit time integration methods may be employed for this purpose.

5.2. Explicit time integration

According to the central difference formulae:

$$\underline{U}_{N+1} = \underline{U}_N + \Delta t_N\,\underline{\dot{U}}_N + \frac{\Delta t_N^2}{2}\,\underline{\ddot{U}}_N \qquad (5.2.)$$

$$\underline{\dot{U}}_{N+1} = \underline{\dot{U}}_N + \frac{\Delta t_N}{2}\,\underline{\ddot{U}}_N + \frac{\Delta t_N}{2}\,\underline{\ddot{U}}_{N+1} \qquad (5.3.)$$

Here, N denotes the time step number, Δt_N being the time increment between t_N and t_{N+1}. If the response is strongly nonlinear, the central difference method should be used with a variable time increment Δt_N for numerical stability. After each time step, a new time increment Δt_N is established from the current stability criterion

$$\Delta t_N = \min\ \{\mu\,L^K \sqrt{\frac{\rho(1+v)(1-2v)}{E(1-v)}}\ |,\ \text{to each element K}\} \qquad (5.4.)$$

where the stability factor μ is taken to be 0.5 for 2D analysis and 0.3 for 3D analysis; L^K is the smallest distance between adjacent nodes of any element K with the same material; E is Young's elastic modulus; v is Poisson's ratio; ρ is the mass density.

Substituting equations (5.2), (5.3) into equation (5.1), lumping both the consistent matrix $\underline{M}$ and the elastic stiffness matrix $\underline{K}_{el}$ and considering Rayleigh-Ritz damping at node J:

$$C^J = \alpha_M\,M^J + \alpha_K\,K_{el}^J\ ,$$

the acceleration at node J in direction x_i can be derived :

$$(\ddot{U}_i^J)_{N+1} = \frac{(F_i^J)_{N+1}}{(M_i^J)_{N+1}} \tag{5.5.}$$

where,

$$(F_i^J)_{N+1} = ((F^{ext})_i^J - (F^{int})_i^J - \frac{\Delta t_N}{2} \{\alpha_M M^J + \alpha_K K_{el}^J\})_{N+1} \tag{5.6.}$$

$$(M_i^J)_{N+1} = (M^J + \frac{\Delta t_N}{2} \{\alpha_M M^J + \alpha_K K_{el}^J\})_{N+1} \tag{5.7.}$$

5.3 Implicit time integration

For implicit time integration, equation (5.1) is usually written as follows :

$$\underline{R}_{N+1} = \underline{M}\,\underline{\ddot{U}}_{N+1} + \underline{C}\,\underline{\dot{U}}_{N+1} - \underline{F}_{N+1}^{ext} + \underline{F}_{N+1}^{int} = 0 \tag{5.8.}$$

where $\underline{R}_{N+1}$ is the so-called out-of-balance force. The Newmark integration scheme consists of following difference formulae:

$$\underline{U}_{N+1} = \underline{U}_N + \Delta t_N\,\underline{\dot{U}}_N + \frac{1}{2}\Delta t_N^2 [(1-2\beta)\,\underline{\ddot{U}}_N + 2\beta\,\underline{\ddot{U}}_{N+1}] \tag{5.9.}$$

$$\underline{\dot{U}}_{N+1} = \underline{\dot{U}}_N + \Delta t_N [(1-\gamma)\,\underline{\ddot{U}}_N + \gamma\,\underline{\ddot{U}}_{N+1}] \tag{5.10.}$$

Substituting (5.9),(5.10) into (5.8), the out-of-balance force becomes an implicit function of $\underline{U}_{N+1}$ only:

$$\underline{R}_{N+1} = \underline{R}(U_{N+1}) = 0 \tag{5.11.}$$

In order to satisfy equation (5.11) for the displacement $\underline{U}_{N+1}$, an equilibrium iteration sequence (p=1,2,...) is required.

Assuming an approximation $\underline{U}_{N+1}^p$ to $\underline{U}_{N+1}$, we admit that in its neighborhood linear mapping:

$$R^{p+1}_{-N+1} = R^p_{-N+1} + K^p_{-N+1} (U^{p+1}_{-N+1} - U^p_{-N+1})$$ (5.12.)

is a good approximation to (5.11), where, K^p_{-N+1} is the tangent matrix:

$$K^p_{-N+1} = \left. \frac{\partial R}{\partial U} \right|_{U^p_{-N+1}}$$ (5.13.)

$$= (\frac{1}{\beta \Delta t_N^2} + \frac{\gamma \alpha_M}{\beta \Delta t_N}) \underline{M} + (1 + \frac{\gamma \alpha_k}{\beta \Delta t_N}) (\underline{K}^{int})^p_{N+1} - (\underline{K}^{ext})^p_{N+1}$$

in which,

$$(\underline{K}^{ext})^p_{N+1} = \left. (\frac{\partial F^{ext}}{\partial U}) \right|_{U^p_{-N+1}}$$ (5.14.)

describes the dependence of the external forces on geometry, and

$$(\underline{K}^{int})^p_{N+1} = \left. (\frac{\partial F^{int}}{\partial U}) \right|_{U^p_{-N+1}}$$ (5.15.)

is the static tangent stiffness matrix which represents the variation of internal forces with displacements.

6. Results and discussions

The theory presented above has been implemented in the finite element code LAGAMINE developed at the M.S.M. department of the University of Liège. Hereafter some results and comparisons with other codes are given.

6.1. Impact of a cylinder

A circular cylindrical copper bar of length 32.4 mm and radius 3.2 mm impacts normally a rigid frictionless wall at 227 m/s (Fig. 6). Axisymmetric analysis is to be performed and the corresponding finite element mesh is 5×50 along the radial and axial directions respectively (Fig. 7). The material properties of copper are: Young's modulus $E = 117$ GPa, Poisson's ratio $v = 0.35$, initial yield stress $\sigma_0 = 400$ MPa, tangent modulus $E_t = 0.1$ GPa, mass density $\rho = 8930$ kg/m³. The results to be obtained for comparison at $t =$

80 μs are: the final deformation and Von-Mises stress shown on Figs. 6 and 7 which are identical to the results given by computer codes DYNA, MARC, NIKE2D;[12] the other results are listed in Table 1. The comparison results are obtained from Ref.12.

In our simulations, two groups of results in Table 1 were computed with 1) explicit scheme; 2) implicit scheme. It is clear that the mixed solid element gives accurate results.

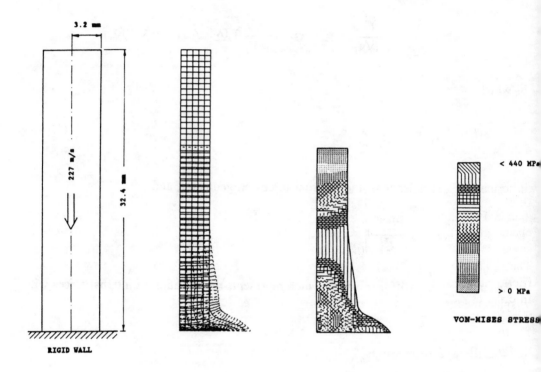

Figure 6 : Data   Figure 7 : Mesh   Figure 8 : Von Mises stress

Code	Final length (mm)	Final mushroom radius (mm)	Max. equivalent plastic strain	Maximum Von Mises stress (MPa)
MARC	21.66	7.023	3.24	476
DYNA2D	21.47	7.127	3.05	472
DYNA3D	21.47	7.034	2.96	-
NIKE2D	21.47	7.068	2.47	-
LAGAMINE 1	21.40	7.157	2.83	477
LAGAMINE 2	21.40	7.118	2.86	472

Table 1. Comparison of MARC, DYNA, NIKE2D AND LAGAMINE

6.2. Impact of die on a solid cylinder

Simple impact in which a solid circular cylinder is axially compressed between two flat rigid dies is a very basic test in metal-forming processes. The workpiece has an initial diameter of 20 mm and height of 30 mm. The constitutive law of the workpiece is elastic-plastic with: Young's modulus $E = 200\ GPa$; Poisson's ratio $v = 0.3$; initial yield stress $\sigma_0 = 0.7\ GPa$; tangent modulus $E_t = 0.3\ GPa$ and mass density $\rho = 7800\ kg/m^3$. The penalty coefficients are: on the contact pressure $K_p = 5 \times 10^5\ MN/m$ and on the shear frictional stress $K_\tau = 5 \times 10^5 MN/m$, and the Coulomb's friction coefficient $\phi = 0.3$ is used. In order to study the impact effects, the computations are carried out for die velocities of $v = 10\ m/s$ and $200\ m/s$. The geometry and finite element model are shown in Figs. 9 and 11. In Figs. 9 and 10 only a quarter of workpiece is considered due to symmetry. Figure 9 illustrates the deformed meshes and Fig. 10 exhibits the contours of equivalent Von-Mises stresses at a stage of 30% height reduction. Figure 11 and Fig. 12 give the deformed meshes and maps of equivalent Von-Mises stresses at different stages of height reduction.

It can be seen that under the low speed impact ($v = 10\ m/s$), the results are quite similar to those of the static solution and 'barrelling' usually occurs due to friction. However, as the die velocity increases: 1) instead of 'barrelling', 'mushshrooming' occurs which coincides with the observation in experiments and in numerical solutions given by other investigators;[2] 2) the maximum value of equivalent stress appears near the interface between workpiece and die instead of near the equator of the cylinder in the cases of static analysis and low imposed speed; 3) contrary to static analysis, the symmetry can not be used, due to wave propagation (see Fig. 12). On the above figures, the agreement between the results given by explicit and implicit methods and by 2D and 3D simulations is excellent.

6.3. Simulation of high speed rolling

Most industrial rolling mills operate at high speed in which case the inertia of workpiece can have a considerable effect on the deformation mechanics.[3] For this example, the deformation is modelled as plane strain and the initial dimensions of the workpiece are 1 m in length and 0.5 m in height, and the diameter of the rolling tool is 2.5 m. The simulation is performed for 20% thickness reduction of the workpiece with surface tangential velocity 100 m/s, and an initial workpiece horizontal speed of $v_s = 20$ *m/s*. The constitutive law of the workpiece is elastic-plastic with $E = 120$ *GPa*; $E_t = 1$ *MPa*; $\sigma_0 = 500$ *MPa*, $v = 0.4$ and mass density is $\rho = 7800$ *kg/m³*. The penalty coefficients on the normal contact pressure K_p and on the frictional shear stress K_t are 5×10^5 *MN/m*, and the Coulomb friction coefficient $\phi = 0.5$ is chosen.

Figures 13 and 14 show the deformed meshes and maps of Von-Mises stresses at several time stages. The sequence of grid distortion clearly reveals the history of deformation of the workpiece. Of particular interest is the substantial rotation at the leading and tail edges of the workpiece. Moreover, the deformation is not uniform across the thickness of the sheet. The material in the layer in contact with the roll experiences the most intense rotation. The material in this contact layer first rotates forward towards the leading edge, then rotates backwards towards the tail edge. These phenomena coincide with other experimental and numerical observations.[3] The residual stresses can be observed on Fig. 14 and they are almost constant along the length of workpiece except for both ends. It is worth noting that the overlapping between workpiece and tool is too small to be observed.

6.4. Analysis of impact-contact of two elastic bars

As shown in Fig. 15, an elastic bar with the initial velocity $v = 0.1$ *cm/s* impacts with another elastic bar at rest. The material properties are taken as YOUNG's modulus $E = 1$ *N/cm²*, POISSON's ratio $v = 0.0$, and density $\rho = 0.01$ *kg/cm³*. For comparison, three simulations are performed :

a)   4-node mixed plane strain elements with implicit scheme;
b)   4-node mixed plane strain elements with explicit scheme;
c)   three-dimensional 20-node isoparametric element with explicit scheme.

Figure 15 shows that all the numerical results practically coincide with the analytical solutions.[16]

7. Conclusions

In this paper, a mixed solid and contact finite elements procedure for nonlinear dynamic analysis has been developed. It is based on an extension of Hu-Washizu variational principle and uses a corotational formulation. The basic conclusions are:

(1)   by mixed solid elements, the spurious zero-energy modes and locking phenomena are avoided; they are computationally very cheap and give accurate results;

(2)   by mixed contact elements, the special corrections of velocity and acceleration on the contact surface are not required when a reasonably small time step is used, and the overlapping between solid and foundation can be controlled;

(3)   for metal-forming processes, when the applied speed is low the dynamic results can be replaced by the static ones; while, in the case of high speed, these results are very different;

(4)   the explicit scheme, when combined with a lumped mass matrix, does not require the solution of a system of equations. All equations are uncoupled. In comparison with the implicit scheme, it may save CPU-time and requires less storage. However, this scheme is conditionally stable;

(5)   the implicit scheme must solve a large system of equations. This task requires a lot of storage and CPU-time. It is observed that, in numerical simulations, the step size can be larger than with the explicit method, without a loss of convergence;

(6)   the results obtained by explicit and implicit time integration schemes are almost the same, irrespective of 2D or 3D meshes. If the finite element meshes are nearly uniform and the applied speed is high, it is better to use the explicit method, otherwise the implicit one is to be preferred.

References

1.   Chaudhary, A.B. & Bathe, K.J. A solution method for static and dynamic analysis of three-dimensional contact problems with friction, *Computers and Structures*, 1986, **24**, 855-873.

2.   Chou, P.C. & Wu, L.W., A dynamic relaxation finite-element method for metal forming processes, *Int. J. Mech. Sci.* , 1986, **28**, 231-250.

3.   Lau, A.C.W., Shivpuri, R. & Chou, PC., An explicit time integration elastic-plastic finite element algorithm for analysis of high speed rolling, *Int. J. Mech Sci.*, 1989, **31**, 483-497.

4.   Lindgren, L.E. & Edberg, J. Explicit versus implicit finite element formulation in simulation of rolling, *Journal of Materials Processing Technology*, 1990, **24**, 85-94.

5.   Kanto, Y. & Yagawa, G., A dynamic contact buckling analysis by the penalty finite element method, *Int. J. Nume. Method Eng.*, 1990, **29**, 755-774.

6.   Cescotto, S. & Zhu, Y.Y., Non-linear dynamic analysis of 2D or 3D metal-forming processes by finite element, *European Conf. on New Advances in Computation Structural Mechanics*, Giens (France), April 1991.

7.   Charlier, R. & Cescotto, S. Modélisation du phénomène de contact unilatéral avec frottement dans un contexte de grandes déformations, *Jounal de Mécanique Théorique et Appliquée*, Special issue, 1988, **7** (1).

8.   Jetteur, Ph. & Cescotto, S., A mixed finite element for the analysis of large inelastic strains, *Int. J. Numer. Method Eng.*, 1991, **31**, 229-239.

9.   Cescotto, S. & Charlier, R., Variational principles for mixed contact elements', Conf. on Unilateral Contact and Dry Friction, La Grande Motte (France), June, 1990.

10.   Cescotto, S., Computer aided methods for improved accuracy of drop forging tools, BRITE Project no RI1B-0218-C(AM), Research report no 2, Dec., 1989.

11.   Belytschko, T, Ong, J., Liu, W.K. & Kennedy, J.M., Hourglass control in linear and nonlinear problems, *Comput. Meths. Appl. Engrg.*, 1984, **43**, 251-276.

12.   Kamoulakos, A., A Simple Benchmark for impact, Bench Mark, 31-35, July, 1990.

13.   Goudreau, G.L. & Hallquist, J.O., Recent developments in large scale finite element lagrangian hydrocode technology, *Comp. Meth. Appl. Mech. Eng.*, 1982, **33**, 725-757.

14.   Hughes, T.J.R., Taylor, R.L., Sackman, J.L., Curnier, A.C. & Kanoknukulchai, W., A finite element method for a class of contact-impact problems, *Comp. Meth. Appl. Mech. Eng.*, 1976, **8**, 249-276.

15.   Wriggers, P., Vu Van, T., Stein, E. Finite element formulation of large deformation impact-contact problems with friction, *Comp. and Structures*, 1990, **37** (3), 319-331.

16.   Timoshenko, S.P. & Goodier, J.N., *Theory of elasticity*, Mc Graw-Hill, New-York, 1970, pp. 409-420.

IMPACT OF A DIE ON A SOLID CYLINDER (V = 10 M/S)

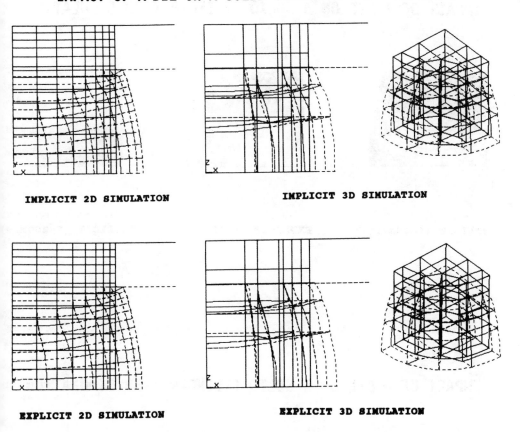

IMPLICIT 2D SIMULATION   IMPLICIT 3D SIMULATION

EXPLICIT 2D SIMULATION   EXPLICIT 3D SIMULATION

IMPACT OF A DIE ON A SOLID CYLINDER (V = 200 M/S)

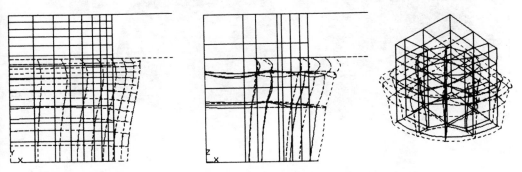

EXPLICIT 2D SIMULATION   EXPLICIT 3D SIMULATION

DEFORMATIONS AT 30% HEIGHT REDUCTION

Figure 9

IMPACT OF A DIE ON A SOLID CYLINDER (2D SIMULATION)

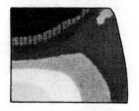 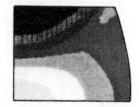 

IMPLICIT (V=10m/s)   EXPLICIT (V=10m/s)   EXPLICIT (V=200m/s)

IMPACT OF A DIE ON A SOLID CYLINDER (3D SIMULATION)

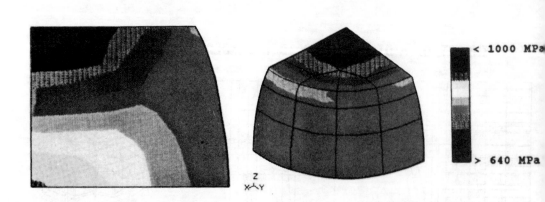

EXPLICIT (V=10m/s)

VON-MISES EQUIVALENT STRESS AT 30% HEIGHT REDUCTION

Figure 10

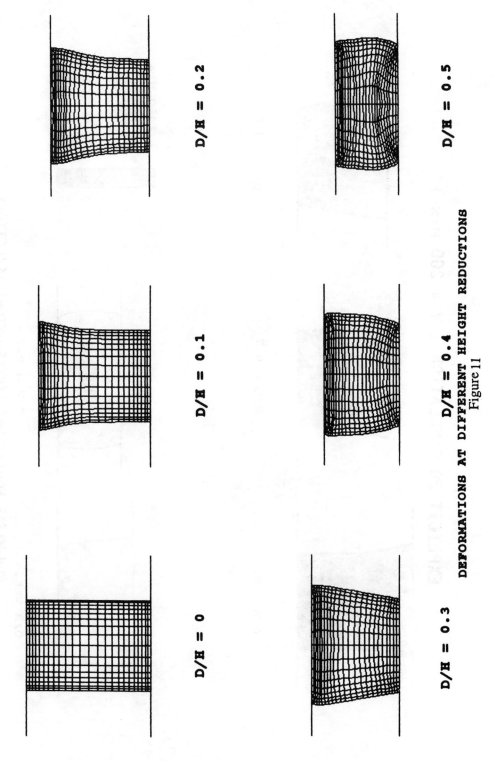

EXPLICIT 2D SIMULATION (V = 200 M/S)

D/H = 0

D/H = 0.1

D/H = 0.2

D/H = 0.3

D/H = 0.4

D/H = 0.5

DEFORMATIONS AT DIFFERENT HEIGHT REDUCTIONS
Figure 11

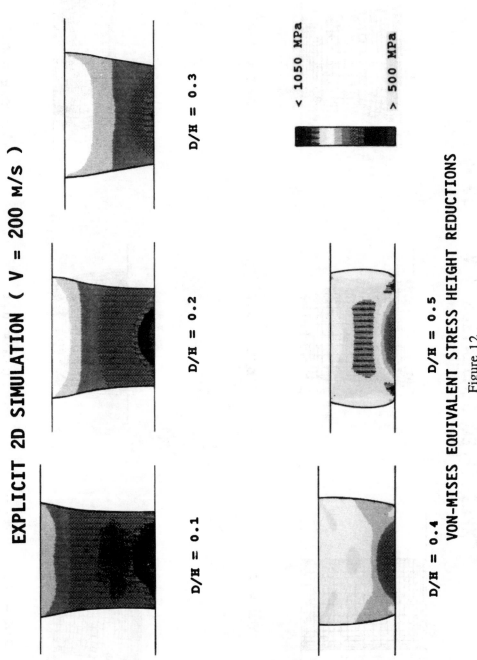

EXPLICIT 2D SIMULATION (V = 200 M/S)

D/H = 0.1   D/H = 0.2   D/H = 0.3

D/H = 0.4   D/H = 0.5

< 1050 MPa

> 500 MPa

VON-MISES EQUIVALENT STRESS HEIGHT REDUCTIONS

Figure 12

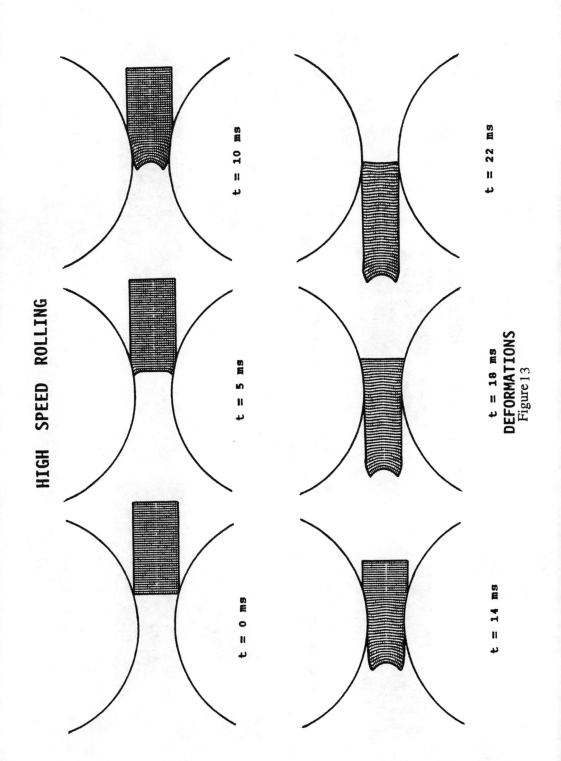

HIGH SPEED ROLLING

t = 0 ms

t = 5 ms

t = 10 ms

t = 14 ms

t = 18 ms

t = 22 ms

DEFORMATIONS
Figure 13

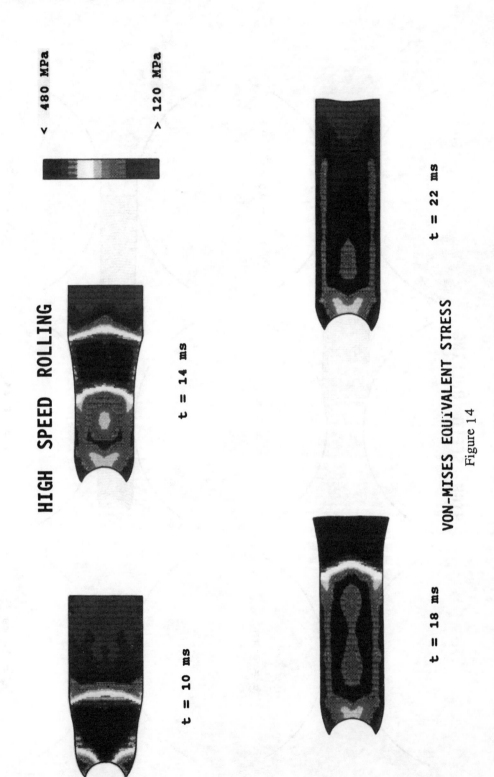

Figure 14

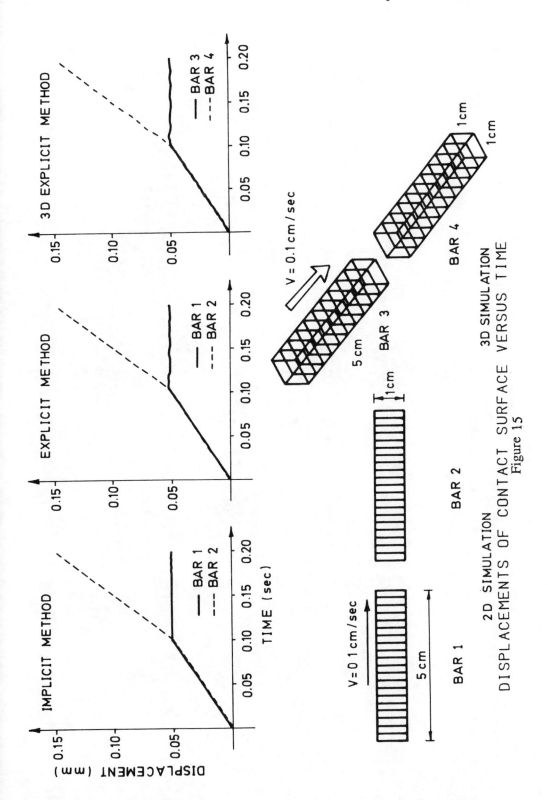

DISPLACEMENTS OF CONTACT SURFACE VERSUS TIME

Figure 15

Chapter 8

Impact analysis of transport flasks

J. Martí

Principia, S.A., Velázquez 94, 28006 Madrid, Spain

ABSTRACT

Engineering and licencing requirements on transport and storage flasks for highly radioactive materials are briefly reviewed from the viewpoint of impact resistance. The various impact problems generated are individually discussed and strategies are proposed for their solution. These concepts are illustrated with examples of impact analyses on actual flasks.

1. INTRODUCTION

Public perception about safety in the nuclear industry has consistently led to higher and higher demands concerning the reliability and performance of equipment and components in that industry. Without entering controversial value judgments, those demands probably do not keep proportion with the corresponding expectations in other areas of human activity. This fact has a number of consequences. On the negative side, an adequate response to such increasing demands has added expenditures, which is one of the reasons why nuclear generating costs have grown rapidly with time. On the positive side, the need to provide adequate responses to safety concerns has generated the motivation and the required financing for scientific and technical developments in many areas. Even in non-specifically nuclear fields, these areas range from seismology and earthquake engineering, which owe a major part of their development to nuclear motivations, to the thermohydraulics of multiphase flows.

Impact analysis has also benefitted from the nuclear industry requirements, as progressively rarer events need to be taken into consideration; also, calculations are always expected to include increasingly greater degrees of complexity and sophistication, as well as a stronger support from experimental verification.

Many sources of postulated impacts exist in the nuclear power generating industry, with some regulatory variations from country to country. The more spectacular ones are external events, such as aircraft impacts on the containment and a variety of other wind-borne or self-carried missiles. Internally, fuel elements, equipment, etc. may be dropped or may vibrate during an earthquake, giving rise to impacts; even pipes are assumed to undergo guillotine breaks and encounter other pipes and equipment during the resulting fast pipewhip motions. The underlying licencing philosophy is that anything that can move must be assumed to do so and to generate impacts under the worst possible circumstances, which then have to be conservatively assessed for safety.

2. IMPACT REQUIREMENTS

The transport and storage of radioactive materials (be those spent fuel, reprocessing waste or other) is an activity which gives rise to a number of potential impact problems. Most national requirements on transport flasks are based on the updated IAEA [1] guidelines. From the viewpoint of impact resistance, they usually demand that the flask sustain successfully all 9m drops onto unyielding plane targets and all 1m drops onto 15cm square punches. All impact attitudes must be taken into account, including the possibility that design singularities (such as lifting trunions, cooling fins or other local features) be directly involved in the impact. The prescribed drop height for testing is obviously related to the allowable speed limits during transport. Thermal and other conditions at impact must be the most pessimistic ones.

It has been mentioned that the flask must sustain "successfully" the prescribed impacts. From a conceptual viewpoint, success is usually assessed on the basis of three different concurrent criteria:

a) No gross structural damage must occur; in other words, the flask must maintain its structural integrity. A corner may flatten, but the flask must not break.

b) The isolation of the contents must also be preserved. This requirement can be stated either in terms of tolerances at the seals or dosage in the proximity of the flask.

c) Damage to the contents must be limited and not lead to thermal, criticality or other problems. This is often replaced by establishing an upper bound on the accelerations felt at the centre of gravity of the flask; the upper bound on the acceleration level must be known to be acceptable for the flask internals.

During transport, a number of existing designs provide the flask with impact limiters or shock absorbing devices at their ends, often involving wood for the shock absorbtion. Others have design features in the flask body which naturally operate in a shock-absorbing capacity because of thinned sections in corner elements, legs or other elements which can dissipate energy by localized plastic straining.

An ideal shock absorber would always generate a constant force, which would obviously be lower than that producing failure on the basis of any one of the three previously stated criteria. The minimum thickness of the shock-absorber would be given by the fact that the generated force, acting over the effective shock-absorbing thickness, must be able to remove all momentum from the impacting flask. In practice, shock absorbers are not ideal and the generated forces are a function of the flask attitude at impact and may also vary with the crushed thickness.

Storage flasks (as well as dual-purpose and multi-purpose flasks, which are used for both transport and storage activities) must also perform successfully when subjected to handling and storage accidents. The transport impact limiters are not in place during storage operations and, as a consequence, flasks which use that protection for transport are somewhat more vulnerable in handling and storage operations. In fact, some storage flasks also include impact limiting devices which are in place during handling and storage.

In any case, the impact requirements which appear during handling and storage operations are different from those associated with transport. They are usually less demanding, but this need not always be the case: it depends on the type of operations to which the flask is to be subjected, in particular, its maximum possible elevation above a resisting surface and the maximum speed at which handling movements take place. The flask must generally be assumed to drop from the highest elevation and in the most damaging attitude, irrespective of the controls and redundancies provided in the lifting and handling processes. Operational moving speeds for the flask are usually of lesser concern, as they are typically very low compared to the velocities attained in the postulated drops.

The impact of flasks has been dealt with by many authors. Continuing sources of information are provided by the PATRAM (Packaging and Transportation of Radioactive Materials) and SMiRT (Structural Mechanics in Reactor Technology) Conferences among others.

3. GOVERNING DEFORMATION MECHANISMS

Flasks are very stiff and competent structures; also, when they drop, regulations usually assign infinite rigidity to the target in an attempt to remain conservative. The assumption that both the flask and the target are simultaneously rigid is obviously not useful for most purposes, since it leads to the prediction of an instantaneous rebound with infinite impact forces and infinite acceleration levels for the flask internals. Hence, a practical assessment of the impact must incorporate some deformability. In principle, any of the three following deformabilities, singly or in combination, can be used in an analysis:

a) The deformability of the shock absorber, whether for transport or storage, if any is in place at the time of impact.

b) The deformability of the flask body and its various features.

c) The deformability of the impacted body.

These deformabilities are briefly discussed in the following paragraphs, since their value governs the momentum transfer during the impact event.

Shock absorber
The deformability of the shock absorber must obviously be taken into account when one is in place. If this is not necessary, a shock absorber should not have been provided in the

first place. For the purpose of calculations, the shock absorber action is often introduced as a quasi-static body force corresponding to the maximum deceleration that the shock absorber would produce on the flask. This acceleration is independently determined by tests and calculations conducted on the shock absorber. However, this strategy has a number of drawbacks:

- Transients are neglected in this approach; they are usually small, but not necessarily negligible.

- Some of the calculated effects are unrealistic. For example, in the course of a base drop, lid bolts may suffer increased loads when the lid tries to rebound before the flask body does. In contrast, the approach mentioned will predict decreased bolt stresses during a base drop.

A second strategy is to replace the shock absorber with an equivalent non-linear force-displacement relationship, which can be acquired either experimentally or by detailed calculations of the shock absorber. The difficulty here stems from the fact that the force-displacement relationship is generally strongly dependent on the flask attitude at impact. Thus, in practice, it is not easy to generate such relationships, let alone to do it to the satisfaction of regulatory authorities.

The third strategy is to model the shock absorber as it is, generally consisting of some arrangement of wood blocks surrounded by a steel liner. The advantage is that few idealizing assumptions are embodied. The drawbacks come from the strong anisotropy of wood and from the very large deformations arising in the impact process; both phenomena are complex and their combination, considerably more so.

In general, the first alternative is the more promising one, although special attention may have to be dedicated to some individual problems (like that mentioned of the lid bolts during base drops) which are not dealt with satisfactorily by this approach. The third strategy is really more adequate for shock absorber design calculations than for analysis of flask survival during impact.

Flask body
The flask body is usually made of one or more types of steel. In certain designs, lead is contained within the steel for radiation shielding purposes.

The behaviour of steel can be described as that of an elastic non-ideally plastic solid, with a von Mises yield surface and an associative flow potential. Strain hardening can be represented with a power-law relationship and strain-rate effects have a very moderate influence on the impact results at the temperatures of interest.

This type of material behaviour is easily represented in any standard finite-element code with non-linear capabilities (ABAQUS [2,3], ANSYS [4], ADINA [5], etc.). If any, problems may arise from lid-flask contacts or because of an inadequate time integration strategy, but not because of the nature of the continuum deformations of the flask body.

Lead does create special problems, often gathered under the term "lead slumping", which is sufficiently descriptive: low strength and high inertia combine to generate pressures acting on the surrounding steel. The rate sensitivity of lead is much more important than that of steel at the temperatures of interest; still, this may not be extremely important if its containment within the steel does not allow too much deviatoric straining. However, a frequent source of difficulties is the contact condition between lead and steel, since a temperature-dependent gap must exist between the two materials due to their different thermal dilation coefficients.

An important consideration in these types of models refers to the values assigned to the properties of the materials involved. Regulations and codes of practice often provide "conservative" figures or material specifications to be satisfied by the material supplier. As will be seen in a later example, in non-linear dynamic analyses it is not always trivial deciding whether the use of an upper or lower bound for a given parameter is actually conservative or not. Often, the only thing that can be stated is that it is not correct; but, regarding conservatism, no definite position can be defended. Hence, the best practice consists on using realistic "best estimates", as opposed to "conservative" values, for the properties involved.

Target

It is evident that unyielding targets only exist in regulations. Furthermore, the flask is a very rigid and competent missile. Thus, if bare flask impacts, the deformations sustained by the target may be as large or even greater than those suffered by the flask. This may have considerable influence in a number of predicted effects, particularly mean decelerations and impact duration.

The undeformability of the target, which is generally taken as the regulatory stance, has several major advantages:

- It is consistent with a site-independent licence for the flask
- It is usually conservative
- It is simple

The latter two items need some additional comments. It is known that, for a given missile and impact velocity, peak contact stresses do not always occur when impacting a rigid target. If the missile maintains a substantial proportion of its velocity after a half-period of the target response, contact stresses in elastic impacts may exceed those arising against an unyielding target. Most flask impacts of interest, though, have contact stresses limited by plastic, not elastic, considerations. For this and other reasons, it remains true to say that the approach is almost always fairly conservative, at least from the viewpoint of flask stresses and deformations, as well as decelerations of the internals.

Simplicity is more important than may initially appear. Incorporating the target deformability is reasonable, and sometimes necessary, when modelling an actual physical experiment. There, the target will deform and the model will err if that effect is left unaccounted. However, in a licencing exercise, accounting for target deformability would require either establishing a "conservative" bound for the response of all possible targets or applying for a double approval, one for the flask and one to ensure that each target will deform as required.

Furthermore, the real target will usually be concrete, in which the deformability occurs by cracking, scabbing and, generally, through processes which are very difficult to model with sufficient accuracy for the intended purposes. And, incidentally, opposing demands are imposed on this concrete: while a soft target will be convenient from the viewpoint of impact considerations, seismic and other structural demands will ensure that it remains fairly stiff and strong.

In summary, target deformability is best left as an existing but unused reserve, unless its utilization becomes absolutely necessary or unless the calculations are intended to represent an actual instrumented experiment.

4. IMPACTS OF INTEREST AND TYPICAL RESULTS

Both engineering logic and nuclear regulations require that all possible attitudes be taken into account when considering flask impacts. During transport, this is generally understood to include every flask orientation. For handling and storage, some orientations may be excluded because of their physical impossibility (e.g. a flask which is only handled at small elevations with its base down cannot possibly involve its lid in the first impact on the ground). It is clear that an infinite number of impact attitudes cannot be analysed. A reasonable approach may include the following analyses (see Figure 1):

- Base drop
- Lid drop
- Corner or edge drops
- Side drop
- Tip-over

The combination of these analyses is generally accepted to provide a conservative bound for all possible effects. The various impacts and their typical consequences are described below.

Base drop
This is the more rigid of all possible impacts, whether or not the flask base is protected with a shock-absorber. Since a large contact area is available from the beginning, the momentum exchange is very rapid and efficient. In this impact, as in all other impacts except that resulting from the tip-over of the flask, the conservative assumption is made

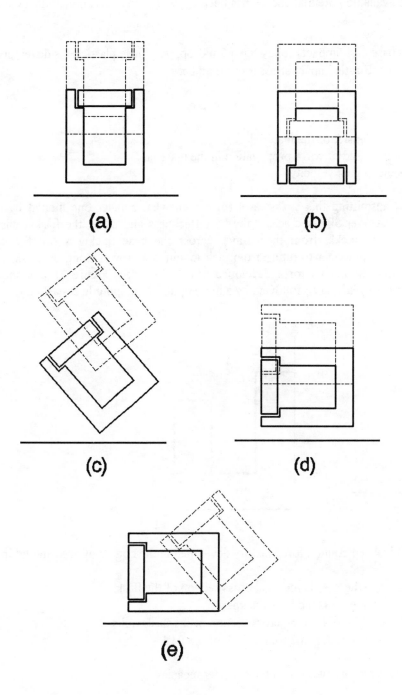

Fig. 1. Impact attitudes of interest for flask: a) base drop; b) lid drop;
c) corner/edge drop; d) side drop; e) tip over impact

that all the available potential energy has been converted to translational kinetic energy at impact.

Even in bare cask impacts following a 9m drop, very little plasticity is developed in the flask materials. Elastic impact stresses σ would be:

$$\sigma = \rho c v$$

where ρ is the density of the material
    c is the speed of wave propagation in the material
    v is the impact velocity

Waves transmitting this stress level (or a somewhat smaller one limited by material yield) travel at over 5km/sec across the base thickness and along the flask walls. These waves rebound quickly from their travel across the base thickness (on the order of 0.1msec), but take longer to return from their longitudinal travel along the flask. In a first approximation, the impact force developed can be idealized (Figure 2) as a short pulse with amplitude F_1 lasting t_1, followed by a longer pulse of amplitude F_2 lasting t_2, where:

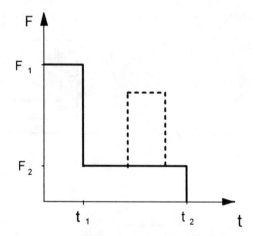

Fig. 2   Idealized contact force history developed in the course of base and lid impacts

- $F_1 = \sigma A_1$, where A_1 is the total cross section of the flask
- $t_1 = 2l/c$, where l is the base thickness
- $F_2 = \sigma A_2$, where A_2 is the cross section of the flask walls
- $t_2 = 2L/c$, where L is the total length of the flask

and the remaining variables retain their former meaning.

The rebound of the base plate leads to a half-cycle oscillation in its natural frequency, during which the base centre loses contact with the impacted plane. If t_2 exceeds that half-cycle duration, the base will impact the plane a second time. This behaviour can be observed in the example provided later (Figure 6).

Base drops are usually noisy impacts; the fast times involved and the quasi-elastic behaviour of the materials contribute to the activation of high-frequency vibrations of the flask structure.

The wave propagation along the walls may be slightly slower than across the base because of the lack of radial confinement. The effects on the lid(s) are analogous to those discussed for the base, except that full oscillation cycles are developed in the case of the lid since there is no target plane arresting the motion after one half-cycle.

It should be noticed that restraining forces must be developed in order to keep the lid from abandoning contact with the flask walls. These forces will initially generate from a decrease in the lid pre-compression (given by bolt pre-tensioning); but, if that is exceeded, the excess would have to be taken as an additional load by the lid bolts, as was already mentioned earlier.

Generally, base drops imply little threat on the flask structural integrity because of the symmetries of the configuration and the large contact area available. However, they are a potential source of problems on three counts:

- The impact duration is minimal (typically 3-4msec in 100ton flasks). This leads to high deceleration values, with whatever implications this may have for the flask internals (possible buckling, etc.).

- The high decelerations also maximise lead slumping effects in flasks which incorporate this material.

- Some attention must be given to lid bolts.

Lid drop
There is no need to discuss this impact in great detail, as the comments would be analogous to those tabled for base drops. The modifications to those comments (e.g. the effects on lid bolts are not relevant in the present case) are fairly obvious and left to the reader.

Corner or edge drops
In these drops, an attempt is made at inflicting maximum damage to the flask structure. The centre of gravity of the flask is assumed to be aligned with the impact point along the line of the flask velocity and the normal to the target plane. In this situation, all of the flask linear momentum will have to be removed by the impact force generated, since no angular momentum is developed during the impact.

The impact point is located on the circumferential perimeter of the base or lid in cylindrical flasks or a corner or edge in cubic or quasi-cubic flasks. Base and lid drops lead to highly symmetric impacts (axial symmetry in cylindrical flasks and at least two planes of symmetry in other flasks). Edge drops typically have a single plane of symmetry and corner drops have at most only one.

The single most important feature of these impacts is the fact that, in bare flask impacts, the initial contact area is infinitesimal. Hence, any finite exchange of momentum must be accompanied by plastic   flattening of the impacted edge/corner. The criteria for assessing the flask acceptability must therefore take account of this inescapable fact.

Plastic straining of the edge/corner leads to a gradual increase of the effective contact area between the flask and the target plane. From the viewpoint of momentum exchange, this is felt as a progressive stiffening of the contact, thus providing a certain shock absorbing capacity. In some designs (e.g. Magnox flask), adequate advantage is taken of this fact by supplying the flask with the equivalent of short deformable legs at the base and the lid.

The contact force history developed in these impacts is approximately made of two straight lines (Figure 3). The first one is associated with the increase in contact area resulting from local flattening. The second straight line represents the elastic rebound following the development of the force peak. This impact is an order of magnitude longer in duration than the corresponding base/lid drops, lasting typically 20-30msec. Because this time is equivalent to several flask periods in its first vibrational mode, the impact is far less dynamic (mobilisation of inertial forces is much smaller) than in the case of base/lid drops. As a consequence, the impact is considerably less noisy and oscillatory deviations from the two straight lines mentioned are only moderate.

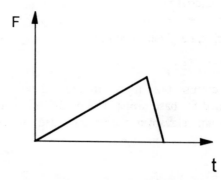

Fig. 3   Idealized contact force history developed in the course of edge and corner impacts

As already indicated, corner/edge impacts are the more demanding ones from the viewpoint of the integrity of the flask, which has to undergo major local deformations. But, although this effect is serious, it is only local; as compensation, this local sacrifice drastically limits deleterious effects elsewhere in the flask.

A potential and undesirable effect of these impacts would be to impose shear loads on the lid bolts. However, flask designs generally provide some "keying" of the lid(s) into the flask body, effectively limiting the magnitude of the shear loads that the bolts may undergo.

Side drop

This refers primarily to impacts on a generator in a cylindrical flask. In bare flask impacts, it has in common with the corner/edge drops the fact that the initial contact area is infinitesimal, which leads to similar shock absorbing effects. However, in the side-drop configuration, the contact area increases much faster as a function of the flask displacement. Hence, the contact stiffens more rapidly and, from the viewpoint of the impact force history, the results appear similar to those of a corner/edge drop with a decreased shock absorbtion capacity.

In many cases (e.g. STC cask), stiffening the lid and base sections of the cylinder or other considerations have resulted in the use of increased diameters in the extreme sections of the flask. In side drops, this has two consequences:

- softening the contact and, therefore, increasing the shock absorbtion capacity
- developing simply-supported beam vibrations in the flask

The latter vibrations entail stresses which combine with other components produced by the impact and its associated stress waves. Also, the rotations imposed on the flask end sections may generate additional forces at the lid bolts; this topic will be further considered in a later section.

The demands generated by side impacts are for the most part comprised between those of the base/lid drops and those of the corner/edge impacts. Structural demands are certainly greater than in the base/lid drops, but smaller than in the corner/edge ones. Conversely, the resulting deceleration levels exceed those arising in corner/edge impacts, without reaching the level of the base/lid cases. If bolts are protected from shearing, their stresses will be primarily those associated with the imposed lid rotations.

The only aspect in which this impact may contribute new structural demands relates to its effects on the flask internals. In spent fuel transport, fuel elements will be located following a direction parallel to the impact plane. This orientation maximises bending of fuel elements, as well as the potential for rupture of fuel rods.

One final consideration is the fact that side and tip-over impacts generate maximum ovalizations of the flask body. The importance of the potential difficulties which may be experienced later when extracting the flask contents by remote control should not be underestimated.

Tip-over

This case is relevant only in storage or handling configurations. Either a small accidental drop during handling or a very considerable seismic action may lead to the overturning of the flask. In this case, the impact occurs with a combination of linear and angular momentum. The potential energy of the flask, when standing in unstable equilibrium with its centre directly above an edge, is converted into kinetic energy; the split between its translational and rotational parts is somewhat dependent on the assumed friction between the flask and the ground.

The tip-over or overturning impact is not very different from the side drop, except that normal velocities are now linearly distributed along the flask, rather than uniform for all points. But, for similar impact velocities, this does not lead to greatly different demands on the flask or its contents.

There is one aspect, though, which deserves special attention: that is the potential for plastic elongation of the lid bolts. It was mentioned earlier that flasks with wider end sections develop beam modes in side impacts; and that this generates strains in the bolts, as they try to impose rotations on the lid(s). In the case of overturning, this effect is reinforced by the fact that, when the impact takes place, the lid has some angular momentum; this momentum must be cancelled by a redistribution of the existing pre-compressions in the lid or, if that is insufficient, by additional loading of the lid bolts. The two effects (arresting the existing rotation and enforcing a new one) are additive and typically range within the same order of magnitude. Thus, out of the various impact configurations, that arising from overturning is the more likely one to result in plastic elongation of the lid bolts and in the opening of a permanent gap at the lid-wall interface.

5. MODELLING STRATEGIES

Although every problem is amenable to limited hand calculations, the analysis of problems such as those discussed here requires the use of advanced finite element tools.

The symmetries discussed earlier allow certain reductions on the mesh size. Even then, the three dimensional meshes required (with the exception of some base/lid impacts) involve thousands (somes time more than 10,000) of elements.

No special problems usually arise for providing adequate constitutive descriptions for the flask materials, at least while wood is not involved. The natural anisotropy and large resulting deformations of this material increase considerably the uncertainties of the model.

The transient character of the problem implies the use of a time marching procedure for integration of the equations of motion. Furthermore, explicit integration procedures based on Lagrangian meshes (ABAQUS/Explicit [3], DYNA3D [6], PAMCRASH [7], etc.) are the superior alternative. The potential advantage of implicit schemes (their unconditional stability) is almost completely lost when the integration time step must in any case be small in order to allow witnessing the wave propagation process; and, moreover, the strong constitutive and geometric non-linearities require a continuous reformulation of the stiffness matrix.

The setting on initial conditions for the impacts implies solving a problem in its own right. The flask must be in a state of equilibrium under the following actions:

- internal pressure
- pre-stressing of bolts
- applicable thermal conditions

Once this problem is solved, all nodes are given their impact velocities and the flask is allowed to interact with a frictionless stonewall representing the unyielding target plane.

A number of masses are not usually modelled explicitly in impact calculations because of their negligible structural contribution: internals, some shielding materials, etc. often fall in this category. Accounting for the inertial contribution of these masses needs case-by-case careful consideration. For example, lumping these masses uniformly with those of the flask structure will decrease the wave propagation velocity in the steel, thus potentially lengthening the impact duration in a base drop and, therefore, underestimating the mean decelerations.

Finally, the lid bolts are of obvious importance in preserving the isolation of the flask contents. However, the great difference in scales between a bolt and a flask make it difficult to extract too much detailed information about bolt behaviour from a global flask model undergoing an impact. Although the evolution of the mean tension at a bolt and other bulk parameters can be reliably produced, the generation of finer details requires detailed analyses at the bolt scale, in which boundary conditions may be supplied from the analysis of the global flask model.

6. EXAMPLES

Some of the concepts given in the previous sections are best illustrated by reference to calculations and experiments conducted for real flasks.

Figure 4 shows the mesh utilized for modelling various impacts in a flask designed for both transport and storage of spent nuclear fuel. Here it appears in its storage configuration, without the impact limiters provided for transport. A detail of the plastic strain levels generated in the course of a corner impact, postulated in the bare flask during storage, can be seen in Figure 5. As it is often the case, large (over 25%) plastic deformations occur locally; but the inner part of the wall, even near the impacted corner, remains essentially elastic.

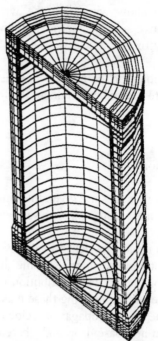

Fig. 4 Mesh used for impact analyses of a transport and storage flask in its storage configuration

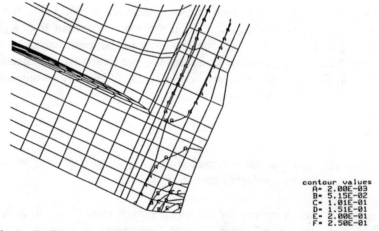

```
contour values
A= 2.00E-03
B= 5.15E-02
C= 1.01E-01
D= 1.51E-01
E= 2.00E-01
F= 2.50E-01
```

Fig.5   Deformations arising in a corner drop during storage operations

The impact force history for a base drop of this same flask appears in Figure 6. The flask impact duration allows for three successive impacts of the oscillating flask base. These three impacts can be clearly identified in the force history; the third one actually occurs when the flask is starting to lose contact with the plane. The fact that the first pulse (compare with F_1 in Figure 2) is decomposed into two is a consequence of a feature of that particular flask base, which is split in two layers by a shielding material. The use of two alternative strategies for dealing with non-structural masses in this impact event showed the potential importance of this decision since considerably different results were generated. Similar types of results have been obtained for other flasks as well as in generic studies of these types of accidents (see, for example, [8]).

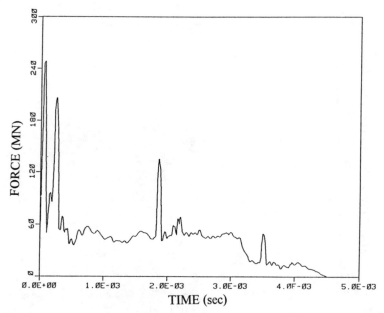

Fig.6   Contact force history during a base drop from a small height without impact limiters

The impact resulting from an overturning or tip-over accident is the cause of the ovalisations shown in Figure 7, which also indicates the opening of a permanent gap at the outer lid by plastic straining of some of the bolts.

Finally, another calculation, also conducted by the author and his colleagues, is shown in Figure 8. Unlike the small heights associated with the previous storage accidents, the drop height in this impact analysis of the Magnox flask is 18m [9]. The calculated and measured curves can be seen to agree very reasonably; also, they approximately correspond to the conceptual scheme illustrated by Figure 3.

7. SUMMARY AND CONCLUSIONS

The consideration of a number of postulated impact accidents is a requirement, both from engineering and regulatory considerations, in relation with flasks designed for transport and/or storage of radioactive materials.

The necessary conservatism can generally be maintained by studying a very limited number of flask attitudes at impact: base, lid, corner/edge, side and tip-over. The aspects in which each one of those impacts may impose the controlling demands have also been discussed in some detail.

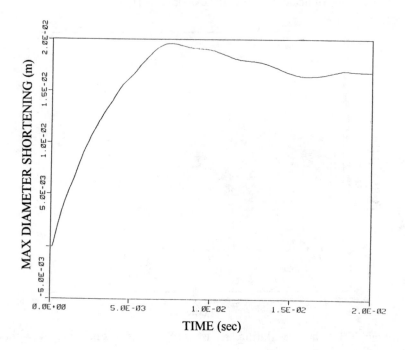

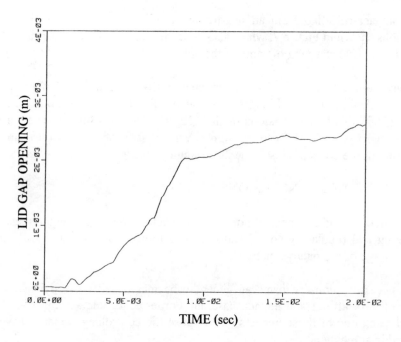

Fig. 7 Ovalization of the flask body and opening of a permanent lid gap following tip-over

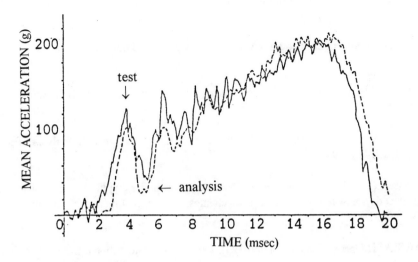

Fig. 8 Contact force history for an 18m corner drop of the Magnox flask

- base/lid and side impacts are worse for the internals
- corner/edge impacts endanger structural integrity
- side drops and tip-over impacts may elongate lid bolts, damage flask internals and compromise their post-impact recovery.

Practical considerations, as well as usual regulatory requirements, make it strongly advisable to neglect the deformability of the target, which is best left as an untapped reserve of conservatism.

Finally, careful attention needs to be paid to bounding assumptions to ensure that they are conservative for each particular criterion under consideration and each specific impact event. This is not always obvious in non-linear dynamic analyses. Examples of potentially dangerous assumptions are the assignment of material properties and the smearing of non-structural masses.

In summary, in spite of the obvious difficulties and cost, it must be concluded that the impact of nuclear industry flasks is a fairly well understood problem and that tools and strategies exist today to solve it with a reasonable degree of accuracy.

REFERENCES

[1] IAEA. Regulations for the safe transport of radioactive materials, International Atomic Energy Agency, Safety Standards no. 6. Viena, 1985 Edition (as amended 1990).

[2] HKS. ABAQUS/Standard. Program Manuals, version 5.3. Hibbitt, Karlsson and Sorensen Inc, 1993.

[3] HKS. ABAQUS/Explicit. Program Manuals, version 5.4. Hibbitt, Karlsson and Sorensen Inc, 1993.

[4] ANSYS Engineering Analysis System. User's Manual, rev. 4.4, Swanson Analysis Systems Inc, 1989.

[5] ADINA. A finite element program of automatic, dynamic, incremental analysis of structures. ADINA R&D Inc, 1991.

[6] Whirley, R.G. and Engelman, B.E. DYNA3D. A non-linear, explicit, three-dimensional finite element code for solid and structural mechanics, Lawrence Livermore National Laboratory, 1993.

[7] PAMCRASH User's Manual, version 12.0, Pam System International S.A., 1992.

[8] Martí, J. Goicolea, J.M., Kunar, R. and Alderson, M. Impact analysis of transport flasks, in Nuclear Containment by D.G. Walton (ed.), pp. 256-266, Cambridge University Press, 1988

[9] Kalsi, G.S. and Dowling, A.R. Three dimensional finite difference analysis of a corner drop of the Magnox nuclear fuel transport flask, 9th International Conference on Structural Mechanics in Reactor Technology, Lausanne, Switzerland, vol. L., pp. 105-112, A.A. Balkema, 1987.